Cambridge

Elements in the Ph
edit
James Owen Weatherall
University of California, Irvine

FROM RANDOMNESS AND ENTROPY TO THE ARROW OF TIME

Lena Zuchowski
University of Bristol

CAMBRIDGE
UNIVERSITY PRESS

Shaftesbury Road, Cambridge CB2 8EA, United Kingdom

One Liberty Plaza, 20th Floor, New York, NY 10006, USA

477 Williamstown Road, Port Melbourne, VIC 3207, Australia

314–321, 3rd Floor, Plot 3, Splendor Forum, Jasola District Centre,
New Delhi – 110025, India

103 Penang Road, #05–06/07, Visioncrest Commercial, Singapore 238467

Cambridge University Press is part of Cambridge University Press & Assessment,
a department of the University of Cambridge.

We share the University's mission to contribute to society through the pursuit of
education, learning and research at the highest international levels of excellence.

www.cambridge.org
Information on this title: www.cambridge.org/9781009500173

DOI: 10.1017/9781009217347

First published 2024

A catalogue record for this publication is available from the British Library.

ISBN 978-1-009-50017-3 Hardback
ISBN 978-1-009-21730-9 Paperback
ISSN 2632-413X (online)
ISSN 2632-4121 (print)

From Randomness and Entropy to the Arrow of Time

Elements in the Philosophy of Physics

DOI: 10.1017/9781009217347
First published online: February 2024

Lena Zuchowski
University of Bristol

Author for correspondence: Lena Zuchowski, lena.zuchowski@bristol.ac.uk

Abstract: The Element reconstructs, analyses and compares different derivational routes to a grounding of the Arrow of Time in entropy. It also evaluates the link between entropy and visible disorder, and the related claim of an alignment of the Arrow of Time with a development from order to visible disorder. The Element identifies three different entropy-groundings for the Arrow of Time: (i) the Empirical Arrow of Time, (ii) the Universal Statistical Arrow of Time, and (iii) the Local Statistical Arrow of Time. The Element will also demonstrate that it is unlikely that high entropy states will always coincide with visible disorder. Therefore, it will dispute that there is a strong link between the Arrow of Time and visible disorder.

Keywords: entropy, randomness, second law, arrow of time, thermodynamics

ISBNs: 9781009500173 (HB), 9781009217309 (PB), 9781009217347 (OC)
ISSNs: 2632-413X (online), 2632-4121 (print)

Contents

1 Introduction

The fact that time is directional appears immediately apparent to us: we experience the effects of this directionality in our own ageing; the fact that we have memories of the past but not of the future; and the way we understand and manipulate causes and effects around us. Philosophers have long tried to understand how to best capture this experiential directionality (e.g., for an anthology of seminal writings, Le Poidevin & MacBeath, 1993), with much of the debate focusing on the question of how to establish a correct ordering of events into past, present, and future.

Furthermore, physicists and philosophers have long tried to find other directional quantities in nature that could be used to 'ground' our experience of the directionality of time. The notion of 'ground' is philosophically contested (e.g., for review, Correia & Schnieder, 2012); in the first instance, it is often unpacked as an 'in virtue of'-relation. Grounding the Arrow of Time in a physical quantity would therefore imply that our experience of time's directionality is in virtue of this quantity and its properties. In this Element, we will further unpack this to mean that the grounding quantity can be used to define and/or explain the grounded one (e.g., roughly following the unpacking of 'grounding' given by Sklar, 1993, pp. 388–96; see also Section 5). In the case of the Arrow of Time, grounding it in a physical quantity means that, ideally, laws pertaining to the quantity can be used to define the directionality of time as we experience it as well as explain why it is experienced this way. Different physical quantities have been put forward as candidates for such groundings (for a comprehensive list, see Roberts, 2022, p. 116): for example, asymmetries in elementary particle decay; the directionality of electromagnetic waves; or the directionality of cosmic expansion. However, one of the oldest and most prominent attempts is one that grounds the Arrow of Time in the thermodynamic or statistical mechanical notion of entropy. Entropy has different formal definitions (Section 3) but has often been positioned as a measure of disorder; therefore, grounding the Arrow of Time in entropy could also establish a connection between time and disorder.

This Element will analyse the grounding of the Arrow of Time in entropy. Thereby, I do not wish to argue that this grounding is superior to any of the alternative proposals mentioned. Rather, the Element aims at:

Aim 1: Reconstructing, analysing, and comparing different derivational routes to a grounding of the Arrow of Time in entropy.

Aim 2: Evaluating the link between entropy and visible disorder, and the related claim of an alignment of the Arrow of Time with a development from order to visible disorder.

In particular, this Element will analyse the roots of the entropy-grounding of the Arrow of Time in the concepts of randomness, entropy, and the second law of thermodynamics. Since there are different co-existing definitions for each of those root-concepts, different versions of this grounding have different roots of different combinations of those concepts. Alternatively, each such root can be viewed as a derivational route to a grounding of the Arrow of Time. This is a novel, uniquely comprehensive approach to analysing the conceptional dependencies of the entropy-groundings of the Arrow of Time and it has several advantages:

(i)   it will allow me to pinpoint the differences between different entropy-groundings of the Arrow of Time by tracing them to differences in their conceptual roots;

(ii) it will allow me to identify epistemic disadvantages and advantages for each derivational route to different entropy-groundings.

(iii) it will allow me to position the derivation of different entropy-groundings of the Arrow of Time in the context of the most prominent controversies in thermodynamics and statistical physics and to show how those controversies have influenced different derivational routes;

(iv) it will allow me to evaluate the relationship between entropy and disorder at the relevant derivational stage and thereby evaluate the posited alignment of the Arrow of Time with disorder.

Advantages (i)–(iv) render this analysis of different entropy-groundings of the Arrow of Time as dependent concepts a perfect framework for the intended aim of this Element series, namely, to provide a unique commentary and introduction to the chosen topic, which will be accessible to postgraduate and higher-level undergraduate students in philosophy of physics. As a matter of fact, this Element is loosely based on a series of six lectures I have contributed to the University of Bristol's Advanced Philosophy of Physics course, which is a credited course for fourth-year undergraduates and MA students but is frequently audited by doctoral students and postdocs as well. However, the Element is not purely a didactic text: I will also demonstrate that the comparative analysis of the different derivational routes to entropy-groundings of the Arrow of Time leads to unique insights and allows one to come to a comparative judgement about the different versions of the groundings that each route makes available (Aim 1, Section 1.1). Furthermore, I will provide a novel evaluation of the claim that statistical entropy is a measure for visible disorder.

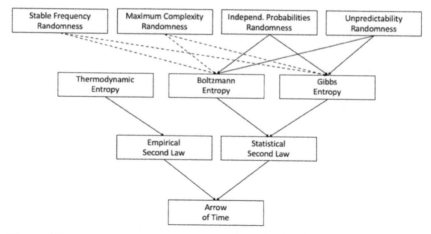

Figure 1 Root-concepts of and derivational routes to entropy-groundings of the Arrow of Time.

1.1 Entropy-Groundings of The Arrow of Time as Dependent Concepts

The first aim of the Element (Aim 1) is to reconstruct, analyse, and compare different derivational routes to entropy-grounding of the Arrow of Time. Aim 1 is based on the premise that entropy-groundings of the Arrow of Time are dependent concepts with root-concepts in the notions of randomness, entropy, and the second law of thermodynamics. This implies the assumption that different derivational routes to different entropy-grounds can be broken down into the subsequent derivation of different versions of the root-concepts. It is impossible, of course, to consider every notion of randomness or entropy ever proposed. However, for each root-concept, I have identified the ones that are of particular conceptual or historical importance. In particular, I will discuss three approaches to defining randomness (Section 2): Process Randomness (Section 2.1); Phenomenological Randomness (Section 2.2); and Inclusive Randomness (Section 2.3). I will then delineate three definitions of entropy (Section 3): Thermodynamic Entropy (Section 3.1), statistical Boltzmann Entropy (Section 3.2.1), and statistical Gibbs Entropy (Section 3.2.2). The three definitions of entropy can be used to derive two definitions of the second law of thermodynamics (Section 4): the empirical second law (Section 4.1) and the statistical second law (Section 4.2). Finally, I will show that the two different second laws can be used to define three different groundings for the direction of time: as foreshadowed earlier, the Empirical Arrow of Time (Section 5.1), the Universal Statistical Arrow of Time (Section 5.2), and the Local Statistical Arrow of Time (Section 5.3). The different root-concepts and their links to concepts along different derivational routes are illustrated in Figure 1. In Section 6, I will present an updated version of this figure, which will include the assumptions we will have identified as crucial for each derivation route in the preceding section.

Using the definition of grounding as sketched out earlier, each entropy-grounding of the Arrow of Time will be shown to fulfil either the definition and/or the explanatory component of a grounding for the direction of time as we experience it. In particular, I will argue that the Empirical Arrow of Time fulfils the definitional component of a grounding for the direction of time (Section 5.1). In contrast, the Universal Statistical Arrow of Time will be shown to have high potential to fulfil both the definitional and explanatory component of such a grounding (Section 5.2), while the Local Arrow of Time has the potential to fulfil the explanatory component (Section 5.3). It will also be demonstrated that the Universal Arrow of Time and the Local Arrow of Time are contingent on several assumptions. The veracity of those assumptions is currently still debated among philosophers and physicists, and an unequivocal judgement of this veracity would require the resolution of long-standing debates in cosmology, neuroscience, and biology. Having gained a nuanced view of the 'state of play' for entropy-groundings of the Arrow of Time, in Section 6, I will sketch out three large-scale conclusions for future research.

1.2 Entropy, Disorder, and the Arrow of Time

A conceptual interpretation of statistical entropy as a measure for visible disorder was established very early on in the history of statistical mechanics (see Haglund, 2017, p. 206 for a summary of its use by Clausius and Boltzmann). It has persisted throughout and has frequently been used to introduce the concept of entropy in physics and philosophy textbooks (e.g., as listed in Haglund, 2017; Rickles, 2016). Viewing entropy as a measure for visible disorder can then also be used to view groundings of the Arrow of Time in entropy as a means of associating time with disorder: if the Arrow of Time points in the direction of increasing entropy, then it also points in the direction of increasing disorder. This association is often used to illustrate the kind of processes we see as indicative of entropy-increasing: for example, by showing the common illustration of a wine glass first being intact and symmetrical and then shattering to become a pile of disordered shards. It is assumed that it is immediately obvious that the visibly disordered state has a higher entropy.

The interpretation of entropy as a measure of visible disorder has two main consequences:

(i) It allows for conceptual arguments asserting that the development of the world (roughly) has to proceed from order to disorder, and that, therefore, evolution and other biological processes can be analysed in terms of order and disorder.

(ii) It establishes a natural connection between entropy and the mathematical concept intended to capture visible disorder, namely randomness.

Consequence (i) is not the primary focus of this Element, but it should be noted that a tendency to explain process in terms of increasing order and (visible) disorder is one of the major 'export products' of statistical mechanics into other academic areas. In particular, there exist several projects that aim to explain biological or societal process as a natural progression from order to disorder (e.g., Gregersen, 2003; Hershey, 2009; Aoki, 2012). For example, in a representative example, Hershey (2009, p. 9) states:

> In simple terms, entropy is a measure of order and disorder, in our human bodies, in so-called inanimate organizations such as corporations, and even the universe. If left alone, these aging systems go spontaneously from low entropy and order, to high entropy and disorder. From life to death, where death is maximum disorder or maximum entropy.

The prevalence of such analyses demonstrates that the interpretation of entropy as a measure of visible disorder is still relevant and of epistemic importance. Accordingly, it provides justification for Aim 2, that is, for a reassessment of the validity of this interpretation. Consequence (ii) has been used in debates on entropy and entropy development in various ways. On the one hand, we will see in this Element that the toy system devised to illustrate the validity of the second law (positing increased entropy development) often deliberately implements a form of process randomness (Section 2.1), which is likely to produce visibly disordered, high-entropy phenomenologies (Section 4.2). On the other hand, the fact that there are natural phenomena which do not seem to match onto any notion of phenomenological randomness (Section 2.2), but still have higher entropy, has been used to argue that the interpretation of entropy as a measure of visible disorder is wrong (Burgers, 1970; Denbigh, 1999; Frenkel, 1999). In particular, there are materials in which the crystalline form has a higher entropy than the (more visibly disordered) liquid one (Frenkel, 1999, pp. 27–8; Gobbo et al., 2020). In Section 3.3, I will argue that it is correct that entropy cannot be viewed as a measure of visible disorder as conceptualised by randomness.

However, statistical entropy can still be interpreted as a measure of abstract disorder, that is, high entropy indicates random distribution over a scenario-dependent phase space and partition. As such, I will maintain that viewing entropy as a measure of disorder is still a useful interpretation but that the relevant notion of order needs to be a more abstract one and cannot be reduced to visible disorder.

2 Three Ways to Define Randomness and Disorder

Consider the following two sequences:

(S1) 001110001000110101100111001101010101010110111001

(S2) 010010000110010101101100011011000110111100100001

Are these random sequences? I think it is fair to say that, at first glance, both sequences look 'visibly disordered', that is, they do not seem to possess easily recognisable patterns. This seems to be particularly apparent in comparison to a clearly patterned sequence:

(S3) 010101010101010101010101010101010101010101010101

However, the two visibly disordered sequences have been generated in very different ways. Sequence (S1) has been generated by performing 48 coin-tosses and recording the results as 0 for 'heads' and 1 for 'tails'.[1] Sequence (S2) has been generated by writing out the phrase 'Hello!', translating it into ASCII characters and then converting the ASCII decimals into binary numbers.[2]

In this section, we will discuss how the notion of 'visibly disordered' relates to different formal definitions of randomness. I have chosen the term 'visibly disordered' to describe the lack of visibly apparent patterns as exemplified by sequences (S1) and (S2), as this is the term most often used in an associated debate on whether high entropy states are associated with visually disordered phenomenologies (Section 3) and whether an increase in entropy always leads to an increase in visual disorder (Section 4). As described in Section 1.2, the notion that an increase in entropy leads to an increase in visible disorder and that, therefore, an entropy-grounded Arrow of Time points in the direction of increasing disorder has been prominent in conceptual debates about the Arrow of Time and in the linking of such debates to other scientific debates, for

---

[1] Sequence (S2) was created by tossing a twenty pence coin, thrown manually and landing on a pad of paper.

[2] The word 'Hello' translates into ASCII code in the following way: H = 72, e = 97, l = 108, o = 111, ! = 33. These numbers in binary notation read: 72(H) = 01001000, 101(e) = 01100101, 108(l) = 01101100, 111(o) = 01101111, 33(!) = 00100001. Therefore, writing out the phrase 'Hello!' in its binary notation, without spaces, results in sequence (S1).

example, on evolution. However, as will be demonstrated in Section 3.4.2, there is no straightforward link between formalised entropy notions and visible disorder as captured by different definitions of randomness. Furthermore, as we will discuss in this section, different definitions of randomness capture different aspects of visible disorder but – so far – none is able to fulfil all the desiderata one might pose for such a formalisation.

When considering sequences (S1) to (S3) and the question of which ones of those we want to call random, the first question we face is whether it matters that the sequences are generated in different ways, that is, is randomness a phenomenological property that can be diagnosed from the sequence alone or a dynamical property that depends on the characteristics of the generating process? We will therefore distinguish three principal ways of defining randomness: process randomness (Section 2.1), phenomenological randomness (Section 2.2), and inclusive randomness (Section 2.3), where both process and phenomenological criteria are included. Within each of those three approaches we face the question of which specific criteria – process and/or phenomenological criteria – we require for a sequence to be called random. As we will see in what follows, different randomness definitions will answer each of those questions differently.

Before discussing different randomness definitions, we should establish a clear criterion against which to judge those definitions. In this, I follow Eagle's (2005) approach. However, as we are particularly interested in using randomness as a way to formalise a notion of disorder, which will then be useful in deciding whether visible disorder is correlated with entropy (Section 3.4), I will focus on one criterion in particular, namely, how well different randomness definitions distinguish between the visibly disordered and visibly ordered sample sequences (S1)–(S3).

With respect to whether each sequence (S1) to (S3) should count as visibly disordered, it is reasonable to assume that we have strong intuitions about (S1) and (S3). It seems clear that sequence (S1), generated by the coin toss, should be defined as random by any good randomness definition. Conversely, it also seems clear that sequence (S3), the obviously patterned sequence, is not random and should be excluded by any good randomness definition. Sequence (S2) looks visibly disordered at first glance but a deeper analysis even of its phenomenology would quickly reveal a pattern, for example, even if the underlying 'code' is not fully deciphered, the fact that the subsequence 01101100 is repeated twice would quickly become apparent. Therefore, given that there is a representation in which (S2) is not random at all, we would like a good randomness definition to exclude (S2) and similar sequences with hidden patterns.[3] This leads us to the first desideratum for a good randomness definition:

R-Desideratum: Good definitions of randomness must include (S1) and exclude (S2) and (S3).

In the following, I will discuss the principal approaches to defining randomness and how they fulfil the R-Desideratum. For each approach, we will outline the most prominent actual definitions: in particular, Independent Probability Randomness (in Section 2.1), Stable Frequency Randomness (Section 2.2.1), Maximum Complexity Randomness (Section 2.2.2), and Unpredictability Randomness (in Section 2.3).

2.1 Process Randomness

Given the importance of the coin toss as a paradigm of generating random sequences, it might appear straightforward to define randomness as simply capturing the specific dynamics we ascribe to this generating process. The crucial feature that renders a sequence random would therefore not be its phenomenology but the process that created it, hence the term process randomness for this category of approaches. The crucial dynamical feature of a coin toss is the fact that each toss is independent of all previous ones and that the probability of obtaining heads or tails will always be 0.5, independent of what results have been obtained in previous tosses. Such a generating process is more generally described as a Bernoulli scheme. Therefore, a randomness definition based on the coin toss as a paradigm of randomness generation could be formalised in the following way:

Independent Probability Randomness: A sequence S is random if it has been generated by a process that can be mapped onto a Bernoulli scheme with approximately equal probabilities.

Auxiliary Bernoulli Scheme Definition: A Bernoulli scheme is a series of independent trials, where for each trial a set of outcomes O_1, O_2, ... O_N can occur with probabilities p_1, p_2, ... p_N and for which $\Sigma p_i = 1$.

---

[3] Eagle (2005) discusses an additional way of generating a visibly disordered sequence: through a chaotic function. The question of whether sequences generated from deterministic, chaotic functions should be viewed as random has been important to philosophers and mathematicians (e.g., Zuchowski, 2012, 2017). However, for establishing a link between entropy and visible disorder, this question is not of particular importance. In particular, Werndl (2009a, 2009b, 2011) has shown that sequences created from chaotic functions are phenomenologically indistinguishable from sequences generated from a coin toss, for example, sequence (S1). Therefore, the majority of the discussion in this Element would simply transfer to a sequence generated from a chaotic function.

The feature that distinguishes a Bernoulli scheme from other probabilistic processes is that each trial is independent of previous ones, for example, in a sequence of coin tosses, the probability of throwing heads (H) or tails (T) is always $p_H = p_T = 0.5$, independent of how many tails or heads have been thrown before. As Kalman (1994, p. 143) notes, this independence of outcomes for each trial is what we look for when designing paradigmatic randomness generating processes (coin tosses, lottery machines, dice, etc). The condition that the probabilities have to be equal has been added to exclude scenarios that intuitively would clearly be seen as not random, for example, a very biased coin toss in which heads occurs with $p_T = 0.99$ and heads with $p_H = 0.01$.

Independent Probability Randomness initially performs well against our desiderata for good randomness definitions. With respect to our **R-Desideratum**, Independent Probability Randomness includes sequence (S1), which – as described earlier – is generated by a process that maps onto a Bernoulli scheme. It also clearly excludes sequences (S2) and (S3), which are not generated by such processes. Independent Probability Randomness therefore performs well against this desideratum and captures our pre-theoretical intuitions about the sample sequences.

Nevertheless, philosophers and mathematicians tend to treat Independent Probability Randomness as either a foil against which to position their own definitions (e.g., Kalman, 1994; Landsman, 2020) or as an outdated definition (e.g., Eagle, 2005). There are two reasons for why Independent Probability Randomness is generally rejected as a suitable definition for randomness: firstly, it excludes sequences generated by deterministic functions (see footnote 3); secondly, there is no guarantee that sequences generated by a Bernoulli scheme actually have visibly disordered phenomenologies. It is the second issue that is particularly important to our discussion about entropy and disorder.

The fact that there is no guarantee that sequences generated by a Bernoulli scheme are similar to (S1) and dissimilar to (S3) can easily be demonstrated by the example of a coin toss sequence: the probability of a sequence of coin tosses producing a sequence of 48 alternating heads and tails, that is, of a coin toss sequence producing (S3), is very small, but not zero.[4] Furthermore, if we performed a much longer series of coin tosses, say millions or billions of tosses, we would actually expect a sequence of 48 alternating tosses to eventually occur naturally with a very high likelihood. Knowing that a sequence has been generated by a Bernoulli scheme is therefore no guarantee that the sequence

---

[4] The probability of any particular sequences of length N resulting from an unbiased coin toss formalised as a two-outcome Bernoulli scheme is $P(N) = 2^{-N}$. Therefore: $P(48) = 4 \times 10^{-15}$.

will fulfil any pre-theoretical or formal notions of phenomenological randomness. Accordingly, Landsman (2020, p. 89) states: 'In other words, their probabilities say little or nothing about the randomness of individual outcomes. Imposing statistical properties helps but is not enough to guarantee randomness [in the sense of visible disorder].'

Given the fact that, for a finite number of iterations, patterned outcomes of a Bernoulli scheme are much less likely than visibly disordered ones, Independent Probability Randomness cannot guarantee that processes that fall under its remit will generate visibly disordered phenomenologies but it can serve as a heuristic to identify processes that are likely to do so. As we will see in Section 4.3, this heuristic function is the one that process randomness, and Independent Probability Randomness, in particular, have assumed in the debate on entropy increases in systems of particles, that is, it has been argued that if the dynamics of a particular system fulfil the Independent Probability Randomness definition, then the system will display highly disordered behaviour and tend towards high entropy states.

2.2 Phenomenological Randomness

A straightforward way of avoiding the possible disconnect between randomness and visible disorder faced by process randomness (Section 2.1) is to define randomness phenomenologically, that is, through criteria that do not refer to the generating process but only to the sequence of outcomes. The general strategy of such phenomenological definitions is to provide a formalisation of the lack of patterns that characterises sequences like (S1). However, a lack of pattern can be captured in different ways. The two most prominent phenomenological randomness definitions are Stable Frequency Randomness (Section 2.2.1), which formalises visible disorder as the inability to identify differences in the frequencies of outcomes throughout a sequence, and Maximum Complexity Randomness (Section 2.2.2), which formalises visible disorder as the inability to use patterns to compress the description of a sequence. As we will see in the following discussion, each of those definitions suffers from some conceptual problems that makes their direct application to the kind of scenarios we encounter in the debate on entropy impossible. Nevertheless, both have been used to approximately describe the visible disorder that traditionally is associated with high entropy.

2.2.1 Stable Frequency Randomness

An influential early class of randomness definitions (e.g., van Mises, 1957) relied on the assumption that an absence of patterns can be captured by requiring that no such patterns can be statistically identified, that is, that there is no possibility of breaking the sequence down into frequently or less frequently occurring outcomes or patterns of outcomes. This can be captured statistically by requiring that any given outcome and pattern of outcomes has the same (stable) probability of occurring as any other one. I will adapt Eagle's (2005, p. 756) formalisation of this definition:

Stable Frequency Randomness:
An infinite sequence S of outcomes of type $O_1 \ldots O_n$ is sf-random if and only if (i) every outcome of type O_i has a well-defined relative frequency f_i^S in S; and (ii) for every infinite subsequence S^* chosen by an admissible place selection, the relative frequency remains the same as in the larger sequence $f_i^{S^*} = f_i^S$.

There is an immediate formal lacuna in this definition: it begs the question of what would count as an admissible place selection, that is, which procedures can be used to select the subsequence S^*, without deliberately enforcing a particular result. For example, a procedure that follows the rule 'an outcome O_n of S is a member of S^* if and only if '$O_n = 1$' will always result in a sequence of all 1s, independent of the set-up of S. This procedure would therefore not be admissible. A number of different place selection mechanisms have been proposed (e.g., for review, Coffa, 1972). Among those, the most prevalent stipulations are the requirement that place selection must be governed by a recursive function (Church, 1940) or recursive statistical sampling (Martin-Loef, 1966).

Setting aside disputes about the correct place selection mechanism, which have no bearing on the main topic of the Element, how does Stable Frequency Randomness fare when measured against our **R-Desideratum** for a good randomness definition? Unfortunately, it is immediately apparent that this definition would not be applicable to any of our examples (S1)–(S3), since all three of those sequences are finite, while – in order to obtain mathematically well-defined frequencies in the definition – Stable Frequency Randomness only applies to infinite sequences. Furthermore, the generation mechanism for two of the sample sequences, (S1) and (S2), that is, coin toss and string to binary translation, are unable to generate infinite sequences under any reasonably realistic conditions, for example, under the stipulation of being given finite time to complete the task. Stable Frequency Randomness therefore does not strictly fulfil our R-Desideratum.

Despite the fact that Stable Frequency Randomness is strictly not applicable to our four test sequences, it will be didactically worthwhile to apply it heuristically and thereby provide the reader with an insight into how the definition is meant to formalise visible disorder as a lack of patterns. Table 1 summarises the relative frequencies of each of the outcomes $O_1 = 1$ and $O_2 = 0$ for each

Table 1 Relative frequencies f_i^S for the two possible outcomes of each of the three sample sequences (S1)–(S3).

	(S1)	**(S2)**	**(S3)**
$O_1 = 1$	0.5	0.5	0.5
$O_2 = 0$	0.5	0.5	0.5

sequence. The task will now be to check whether it is possible to recursively pick a subsequence which has different frequencies of those outcomes.

This is easily done for the patterned sequence (S3): let's define the subsequence as $S3^*$: $S3^*{}_n = S3_{2\,n}$, namely, the nth element of $S3^*$ is the $2n$th element of $S3$, or, in other words, by taking every second element of $S3$. This results in the subsequence:

(S3*) 111111111111111111111111

The relative frequencies of (S3*) are: $f_1^{S3^*} = 1, f_2^{S3^*} = 0$, namely, different from the ones of the parent sequence (S3). Accordingly, in line with our pre-theoretical intuitions, the patterned sequence (S3) is not Stable Frequency Random. As a matter of fact, in this case, we can easily see that this would be true even if both the sequence and the subsequence would be extended infinitely, so that the definition would be formally applicable to the sequence.

What about the hidden-pattern sequence (S2)? Let's assume we have the ability to exhaustively scan the sequence for recurring subsequences; something a computer could easily do, even if it would be exhausting and time-consuming for a human. Such a scan would show that there is one subsequence, $S2^{max}$, which can unequivocally be identified as occurring most frequently in $S2$, namely the one that corresponds to the binary translation of 'l' in 'Hello!':

(S2^{max}) 01101100

We can now define a second subsequence $S2^*$ recursively by stipulating that $S*_n = 1$ if S_n is part of $S2^{max}$ and $S^*n = 0$ otherwise. This leads to:

(S2*) 000000000000000011111111111111110000000000000000,

with relative frequencies $f_1^{S2^*} = 1/3, f_2^{S2^*} = 2/3$, which are different from the ones of the parent sequence (Table 1). Accordingly, (S2) is not Stable Frequency Random. It is noteworthy that the preceding analysis is entirely phenomenological: identifying the most frequently occurring subsequence does not require us to know that it corresponds to the binary translation of the symbol 'l'. The detection of the hidden pattern is therefore dependent on the fact that the English language has certain regularities, namely, that certain letters and combination of letters occur with a higher frequency than others. We would also expect those regularities to continue to exist in any string of proper English words; thus we can be confident that, even in the infinite limit to which Stable Frequency Randomness is actually applicable, we would obtain the same result. However, there are examples of hidden pattern sequences which do not have those regularities and would therefore be judged to be Stable Frequency Random. Landsman (2020, p. 89) discusses the example of Champernowne's number, which is formed by appending all natural numbers to each other:

(S-C) 012345678910111213 . . .

It can be shown that this sequence is Stable Frequency Random, even though it clearly has a hidden pattern. Stable Frequency Randomness therefore seems to concur with our intuitions about what is a patterned sequence in cases where the regularities can be detected through a frequency analysis, but not in cases where patterns do not lead to particular differences in the frequency of outcomes.

The heuristic application of the Stable Frequency Randomness definition has thus been successful in excluding the patterned sequences (S2) and (S3). Would it be successful in including sequence (S1), generated by the paradigm-random process of a coin toss? This question highlights an immediate problem with Stable Frequency Randomness: a successful application of the definition requires us to prove a negative, namely that there exists no properly selected subsequence with different relative frequencies than the parent sequence. In contrast to the two patterned sequences, (S2) and (S3), sequence (S1) does not provide us with an 'obvious' mechanism for selecting subsequences with different relative frequencies to the parent sequences, and we have no reason to assume that even an exhaustive, automated parsing would lead to the identification of maximally or minimally occurring subsequences. However, while it is possible to do such an exhaustive parsing for a finite subsequence, the case to which Stable Frequency Randomness is actually applicable, that of infinite sequences, would not allow such a parsing to ever reach a conclusion. The heuristic application of Stable Frequency Randomness to our test sequences therefore illustrates that this definition of randomness appears to work best as a methodology for identifying and excluding patterned sequences, rather than as one to formally identify visible disorder.

2.2.2 Maximum Complexity Randomness

The second class of prominent phenomenological randomness definitions is based on the idea that such definitions should capture the lack of patterns we pre-theoretically ascribe to the concept more directly, and without having to

solve the thorny issue of admissible place selection. Furthermore, these definitions also tie in with two mathematical concepts that became popular in the
1960s, namely computability and compressibility. With respect to sequences,
computability means that there exists an algorithm to reproduce a given
sequence, that is, there exists a programme that can be run on a computer.
Compressibility is defined as the existence of an algorithm that is shorter
(according to some measure, see the following discussion) than the sequence
itself.

This class of definitions then specifies that randomness is equivalent to incom-
pressibility (e.g., Kolmogorov & Uspenskii, 1988), that is, that it is not possible to
use patterns in the sequence to device an algorithm that reproduces it and is
shorter than the sequence itself. Rather than using the term incompressible, such
definitions usually refer to complexity, which is a notion that more easily admits
degrees. A sequence is more complex the less it can be compressed, that is, the
closer the shortest possible algorithm is in length to the sequence itself. A formal
definition can be given as (adapted from Eagle, 2005, p. 759):

Maximum Complexity Randomness: A sequence S is random if its
complexity is equal or greater than its length, namely, $C(S) \geq l(S)$.

Auxiliary Complexity Definition: The complexity $C(S)$ of a sequence is the
length of the shortest programme P of some Turing machine T which produces

1-Machine

Set of states Q	Q = {A}
Set of symbols S	S = {0, 1}
Initial state q_0	q_0 = A
Transition function T(Q,S) (quintuple formalisation)	;A01RA; ;A11RA;

10-Machine

Set of states Q	Q = {A,B}
Set of symbols S	S = {0, 1}
Initial state q_0	q_0 = A
Transition function T(Q,S) (quintuple formalisation)	;A01RB; ;A11RB; ;B00RA; ;B10RA;

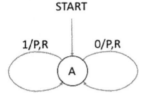

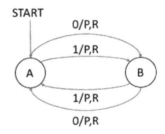

Figure 2 The 10-Turing Machine that reproduces sequence (S3) in comparison
to a 1-Turing machine that prints a sequence of 1s.

S as an output, when given as input the length of *S*. (*C(S)* is set to *l(S)* if there exists no such programme *P*.)

The definition needs to be grounded in a particular class of Turing machines. Turing machines are conceptually minimal computers consisting of a tape, which is divided into sections into which a single symbol (usually 0 or 1) can be printed; a moving head that can read, erase, and print symbols on the tape; and a set of instructions specifying conditional sequences of the MOVE, READ, ERASE, and PRINT operations. Despite the minimal set-up, Turing machines can implement virtually any conceivable mathematical algorithm (for more on the mathematics and capabilities of Turing machines, see e.g. Turing, 1936; Copeland & Proudfoot, 2005).

In the Auxiliary Complexity Definition, it is usually assumed that one would stipulate that the Turing machine *T* on which the algorithms will be implemented is a universal, minimal Turing machine, namely, a Turing machine that can emulate all other Turing machines and uses the most minimal set of symbols and operations to do so. However, there are different ways of constructing minimal Turing machines and no one machine has been designated as 'the minimal, universal Turing machine' yet (for review, see e.g. Margenstern, 2000).

Given those technical preconditions, how does Maximum Complexity Randomness perform against our **R-Desideratum** for a good randomness definition? It is notable that – in contrast to Stable Frequency Randomness – Maximum Complexity Randomness is actually applicable to our four finite sample sequences. However, any mathematically accurate application of the definition would require the construction of several complicated Turing machines, a task that would leave little scope in the Element to discuss anything else. In the following, I will therefore just sketch out conceptually how Maximum Complexity Randomness could be applied to (S1)–(S3) and will indicate what the likely outcome of such an application would be.

It is relatively easy to determine (Figure 2), that the patterned sequence (S3) can be reproduced by a Turing machine with instructions that look roughly like this:

(T1) STATE A: PRINT 1, MOVE RIGHT, GO TO B;
       STATE B: PRINT 0, MOVE RIGHT, GO TO A

There are various different ways of translating Turing machine instructions into binary sequences (e.g., DeMol, 2021). If we use my old trick of translating into ASCII code first and then translating the corresponding numbers into binary code, we need eight binary digits per symbol (including punctuation), resulting in a string of 376 binary digits. That is, of course, considerably longer than sequence (S3) itself. Even when we factor in a cleverer way of translating instructions to binary, it is unlikely that the resulting length of the programme will be much shorter than 48 symbols, that is, the length of (S3). Strictly speaking, (S3) could therefore be judged Maximum Complexity Random, which does not capture our pre-theoretical intuitions and does not fulfil the R-Desideratum. However, it is noteworthy that the set of instructions (T1) can be used to produce a string of any length with the same pattern. If we therefore extended (S3) to a length of, say, 400 symbols, the resulting sequence would not be Maximum Complexity Random, as the programme to reproduce it would still only require 376 (or thereabout) digits. This highlights a problem with Maximum Complexity Randomness: while the definition is formally applicable to finite sequences, it only works well as a measure of compressibility for relatively long sequences. This is due to the fact that while an algorithm might use regularities to achieve a degree of compression, any programme will have some length. Short sequences and single events will therefore almost by default be found to be Maximum Complexity Random.

What about the coin toss sequence (S1)? There is no obvious pattern in sequence we can use to specify instructions similar to (T1) and we have as yet not discovered an algorithm that allows us to reproduce (simulate) the physics of a given coin toss. Accordingly, the only way of reproducing this sequence on a Turing machine would be to ask it to write out the sequence directly, symbol for symbol:

(T2a) STATE A: PRINT 0, MOVE RIGHT, PRINT 1, MOVE RIGHT, PRINT
       0, MOVE RIGHT, PRINT 0 ... MOVE RIGHT, PRINT 0, MOVE
       RIGHT, PRINT 1;

There is an obvious way of making this programme a bit more efficient: it is possible to write a Turing machine programme that copies a given sequence from an input section of the tape and then prints it on a different output section of the tape (e.g., deMol, 2021). This would allow us to pull out all the different PRINT and MOVE RIGHT instructions and set them as one block of code, which is independent of the length of the specified sequence, namely, as a separate COPY instruction, which only needs to be specified once.

(T2b) INPUT: 0100100001100101011011000110110001101111100100001
       STATE 1: COPY INPUT;

The advantage of this minimal formalisation is that we can immediately see that the length of the programme T2 will always be the lengths of the sequence plus whatever symbols are required to specify the COPY instructions in (T2b). Accordingly, the coin toss sequence (S1) is Maximum Complexity Random.

With some caveats concerning the lengths of the sequences, Maximum Complexity Randomness performed well on the two most paradigmatic sample sequences. However, it is considerably more difficult to determine, even conceptually, what would be the likely form of shortest algorithm to reproduce sequence (S2). With respect to sequence (S2), the hidden pattern is one that depends on recognising the word 'Hello' as a proper English phrase. The algorithm would thus have to enable the computer to do this, for example, by providing it with a numbered dictionary of English words. It would then have to have a block of code to perform the translation into binary numbers, for example, by additionally being given an ASCII translation table as input. Conceptually the problem would look something like this:

(T3) INPUT: NUMBER(S) OF WORD(S) TO TRANSLATE.
      SUBPROGRAMME 1: DICTIONARY, WITH RETRIEVAL CODE.
      SUBPROGRAMME 2: TRANSLATION ALGORITHM, WITH ASCII TABLE.
      SUBPROGRAMME 3: PRINT TRANSLATED STRING.

As in the case of the simply patterned sequence (S3), for a short sequence, this programme will surely be longer than the length of the sequence itself. However, the three subprogrammes in (T3) are again constant in length, such that the length of the programme will remain constant even if we ask it, say, to translate a long string of English words one after the other. As long as the dictionary labelling is done in a way that the label is usually shorter than the word itself, the programme to translate a long sequence of words would eventually be shorter than the sequence itself, therefore rendering it not Maximum Complexity Random. Those conceptual considerations might be enough to convince ourselves that Maximum Complexity Randomness does fulfil the R-Desideratum, for example, that, for relatively long sequences at least, it matches our intuitions that sequence (S1) is random while sequences (S2) and (S4) are not. However, due to the multi-realisability on different Turing machines and the fact that it is only applicable to relatively long sequences, its potential for practical application is limited.

2.3 Inclusive Randomness Definitions

Following his review and rejection of both process randomness (Section 2.1) and phenomenological randomness (Section 2.2), Eagle (2005, p. 755) presents a randomness definition that has been devised specifically to meet his desiderata:

Unpredictability Randomness: An event E is random for a predictor P using theory T if and only if E is maximally unpredictable.

Auxiliary Unpredictability Definition: An event E is maximally unpredictable for P and T if and only if the posterior probability of E yielded by the prediction functions that T makes available, conditional on current evidence, is equal to the prior probability of E

Unpredictability Randomness is an inclusive randomness definition, as it allows predictors P to be derived from any theory about the sequence under consideration, including theories about the generating mechanism (as utilised in process randomness, Section 2.1) or any patterns in the actual sequence (as utilised in phenomenological randomness, Section 2.2). It therefore combines process and phenomenological randomness. Eagle (2005) then demonstrates that Unpredictability Randomness indeed fulfils his specific desiderata, namely, that it is applicable to finite sequences and single events; that Unpredictability Randomness can be tested for by statistical tests and sampling; that it allows empirical confirmation and can serve as an explanation for observed behaviour; and that it is compatible with determinism. This argument can be found in the original paper and, I think, is generally successful; maybe unsurprisingly so, given that Unpredictability Randomness was designed specifically to fulfil those desiderata.

However, those advantages come at the epistemic cost of introducing a subjective element into the randomness definition: Unpredictability Randomness is defined relative to a predictor P and their ability to derive predictor functions from a specific theory T. Eagle (2005, p. 779) further specifies that '[i]t is essential to note that judgements of predictability will typically be made by an epistemic or scientific community and not a particular individual'. Within the confines of Eagle's (2005) own project of devising a randomness definition that best captures the use of the concept by scientists, indexing randomness to different scientific communities might not be worrisome. However, when considering randomness as a universally applicable mathematical concept, worries remain about how unequivocally judgements about the Unpredictability Randomness of a given sequence can be made.

This becomes apparent when we consider whether Unpredictability Randomness fulfils our **R-Desideratum**, namely, whether it captures which of the sample sequences we consider to be visibly disordered. For sequence (S1), we have seen earlier that neither our best theories about the physics of the generating coin tosses (Section 2.1) nor any frequency (Section 2.2.1) or compressibility (Section 2.2.2) analyses of the sequence itself make available any predictors that would raise our ability to predict the next coin toss with more than 50/50 accuracy. Accordingly, (S1) is Unpredictability Random. At the current time, there is no scientific disagreement about the fact that the best theories of the physics of coin tosses are not precise enough to lead to any efficient predictors. However, it is not inconceivable that at some point in the future better theories and better computational resources might become available. At such a time, we would have to revise our judgement about the Unpredictability Randomness of sequence (S1).

For sequence (S3), a predictor P is easily available from any theory that recognises the alternating pattern of 0s and 1s in the sequence. Accordingly, (S3) is not Unpredictability Random. Similarly, the next element in sequence (S2) will be easy to predict for anyone who has 'cracked the cipher' and recognised the underlying 'Hello!'-phrase. It is reasonable to assume that most competent English speakers will be able to do so. However, even someone who is not competent in English could avail themselves of the kind of frequency analysis we have conducted in Section 2.2.1 and raise their predictors above the level of chance. Sequence (S2) is therefore not Unpredictability Random. This example does illustrate, however, that under certain conditions, Unpredictability Randomness could become a very subjective notion: imagine a language that has no letter regularities, and which only has a very small community of competent speakers. A sequence encoding a phrase of this language would then be Unpredictability Random to anyone but this very small community of competent speakers. For the given sample sequences, however, Unpredictability Randomness fulfils R-Desideratum 1 and matches our intuitions about their visible disordered appearances.

3 Two Approaches to Entropy

There are two general approaches to entropy: Thermodynamic Entropy (Section 3.1), which treats entropy as a macroscopic property of thermodynamic systems, that is, systems characterised by macroscopic variables like pressure p, volume V, and temperature T; and statistical entropy (Section 3.2), which treats entropy as a macroscopic property of systems best characterised by microscopic variables like particle position x and momentum p. The standard sample system to illustrate the difference between those two approaches is to consider a quantity of gas in a box (or another container): the Thermodynamic Entropy can be computed from the macroscopic variables of the gas considered as one substance, that is, the volume of the box, the gas's temperature, and the pressure the gas exerts on the container; in contrast, statistical entropies will be computed from the properties of the individual particles in the gas, for example, their positions, velocities, and directions.

This Element does not aim to present a chronological history of the development of these two approaches to entropy and their associated specific definitions. Instead, I will highlight the major conceptual issues and debates as they relate to Aim 1 (Section 1.1) and Aim 2 (Section 1.2) of the Element. As such, I will only discuss the three most prominent definitions of entropy: the standard formulation of Thermodynamic Entropy (Section 3.1), which is also the historically earliest; the Boltzmann Entropy (Section 3.2.1), which is a statistical entropy based directly on the microscopic properties of a collection of particles; and the Gibbs Entropy, which is a statistical entropy based on the performing statistics over a virtual ensemble of particle systems (Section 3.2.2).

An influential debate about the merits of different statistical entropies centres around the question of how well each of those entropies reduces the Thermodynamic Entropy. Reduction is a contested philosophical notion, and we will not have space to go into the different interpretations of this concept (for review, see, e.g., van Riel & van Gulick, 2019). In a broad sense, reduction means that a more fundamental notion (in this case, statistical entropy) can be used to fully derive all properties of a less fundamental notion (in this case, Thermodynamic Entropy). Such a derivation will usually involve the formulation of bridging relations between properties used in the two properties. The reduction of Thermodynamic Entropy to statistical entropy is part of a larger project to reduce thermodynamic to statistical physics, that is, to recover all of the laws and theorems formulated in terms of thermodynamic macroscopic variables in a set of laws formulated in terms of microscopic particle properties (with the help of appropriate bridging definitions). The question of how well each of the two statistical entropies performs in the context of this project of reduction directly pertains to derivations of entropy-groundings of the Arrow of Time, which can also involve making assumptions about the reduction of certain processes (Section 5.3). Accordingly, this aspect is directly relevant to Aim 1, tracing out different derivational routes and their contingencies to entropy-groundings of the Arrow of Time. As I will show in Section 3.3, neither Boltzmann nor Gibbs Entropy straightforwardly reduces the Thermodynamic Entropy. However, Boltzmann Entropy offers a more straightforward approach to potential reductions, as it is formulated in terms of particle properties rather than ensemble statistics.

The second aim (Aim 2) of this Element is to evaluate the link between statistical entropy (and thereby entropy-groundings) and disorder (Section 1.2). We have seen that visible disorder is formalised in different ways by different randomness definitions (Section 2). Therefore, we are now in a position to evaluate whether the two statistical entropy definitions map high entropy states onto states with high visible disorder. In Section 3.4, I will show that this is only the case for a very restricted class of non-interacting, elastic particle systems.

I will present a counterexample of a relatively simple system of rod-like particles in which the visibly ordered, crystalline state has a higher entropy than the visibly disordered, liquid state. Therefore, it seems that the link between entropy and visible disorder is not strong. However, I will argue that one could maintain that there is a link between a physically motivated ordering over appropriately defined phase spaces (and partitions) and entropy. The interpretation of entropy as a measure of disorder therefore does not have to be fully abandoned but needs to be supplemented by the recognition that what counts as order and disorder is determined by the specific physical situation and does not straightforwardly map onto notions of visible order and disorder.

3.1 Thermodynamic Entropy

The earliest conceptualisation of entropy took place in the context of nineteenth-century thermodynamics (e.g., Frigg & Werndl, 2011, p. 2). In this early context, entropy was not conceived as a measure of disorder but as a conservation variable that expresses constraints on the heat exchanges during reversible and non-reversible thermodynamic processes. These early formalisations therefore made no reference to individual particles but are entirely based on the standard macroscopic thermodynamical variables, namely, pressure p, volume V, temperature T, and energy in the form of heat Q:

Thermodynamic Entropy: $S_{TD}(B) = S_{TD}(A) + \int\limits_{A}^{B} \frac{dQ}{T},$

where A and B describe different (p, V, T)-states of the system, and S_{TD} is the symbol we will use for Thermodynamic Entropy. An obvious conceptual interpretation of the Thermodynamic Entropy is that the entropy S_{TD} provides a measure of how much energy needs to be put into a system to take it from State A to State B if the temperature T remains constant. The unit assigned to entropy in this early, thermodynamical conceptualisation is therefore [J/K], namely, [energy]/[temperature].

One important consequence of the definition of Thermodynamic Entropy is that the absolute value of $S_{TD}(B)$ is underdetermined, that is, we need to fix an initial entropy value $S_{TD}(A)$ to obtain more than an entropy difference between the states A and B. This is usually done by fixing a zero-entropy state, that is, setting $S_{TD}(A) = 0$. Furthermore, the definition implies that for a closed system, that is, without exchanges of heat Q, the Thermodynamic Entropy is a constant of the system. Conversely, if two states A and B have the same entropy, no heat can be added or subtracted from the system during a transition between those two states. As mentioned earlier, this initial conceptualisation therefore serves as a formalisation on constraints on the possible (p, V, T)-values of a thermodynamical system that can be reached with or without heat exchanges.

3.2 Statistical Entropies

The definition of statistical entropies is routed in the recognition that thermodynamic systems can be described on two levels: as a macroscopic substance (gas, liquid, solid) characterised by macroscopic variables like temperature T, pressure p, and volume V, and as a collection of particles, each characterised by microscopic variables like position x and momentum p. There is a large array of microscopic variables that can be taken into account; beyond position and momentum, particles can also have rotational momentum, spin, charges, and other variables. This variability in what is physically relevant to characterise the microstate of a system, namely, which micro-variables are taken into account, will be important in Section 3.4, when we discuss the link between visible disorder and statistical entropy.

The degrees of freedom of all micro-variables can be envisioned as an abstract space, similar to the spatial one described by the traditional three-dimension coordinate systems for the spatial location variable x. This space is called phase space. In contrast to spatial space, phase space does not just have axes on which the three spatial degrees of freedom can be specified but contains a dimension for each micro-variable's degree of freedom. For example, a phase space formed by the position and momentum variables would be six-dimensional and each particle's microstate could be completely described by the corresponding (x, y, z, p_x, p_y, p_z)-phase space vector. The microstate of a system of particles is therefore specified by the collection of their phase space vectors.

Statistical entropies are formulated in particular phase spaces, usually – but not always (Section 3.4) – the six-dimensional position-momentum space. It is assumed that the particles in the collection under consideration are indistinguishable in all other aspects besides their phase space vectors. Therefore, it is possible to define macrostates as determining the state of the collection particles without reference to individual phase state vectors: for example, statements like 'half of the particles are in each side of a box'; 'all of the particles are in one side of the box'; 'all of the particles have momentums that fall within a given region of phase space'. In this example, we can also notice that – unless we develop a probability measure for the distribution of microstate vectors through phase state and integrate over this (Section 3.2.2) – defining macrostates for a given particle system involves coarse-graining: we have to divide phase space into partitions (halves, quarters, otherwise specified divisions) to define the macro-states as done earlier. Those partitions are not a priori fixed and can be coarser (e.g., division in half) or finer (e.g., division in quarters) for a given system. What it means to choose a 'good' partition has been part of the debate on the merits and properties of different entropies; I will return to this question both when discussing the relation of different entropies to visible disorder (Section 3.4) and – to a lesser degree – when discussing the relationship between statistical entropies and Thermodynamic Entropy (Section 3.3).

The general idea behind statistical entropies therefore is to provide a measure for how microstates relate to macrostates for a given collection of indistinguish-able particles.

3.2.1 Boltzmann Entropy

As indicated by its name, the definition of Boltzmann Entropy originates in Boltzmann (1872). Boltzmann Entropy S_B is defined as a proportionality meas-ure of the number W of microstates that correspond to a given macrostate:

$$S_B \propto W. \tag{3.1}$$

This concept can best be demonstrated by considering a very simple system of four identical, non-interacting particles in a box (Figure 3). Let's assume that we have labelled the particles consecutively from 1 to 4, and that we only consider the positions of the particles in the box as the relative variable. Furthermore, we partition the box into two halves, Half A and Half B. The phase space vector x_i of each particle then just contains one variable, namely the specification whether it is in Half A or Half B. A full description of the microstate of the system is then given by the specification of all four phase space vectors $x_1 \ldots x_4$. In contrast, a description of the macrostate of the system should make no reference to individual phase space vectors and will consist of descriptions like 'four (all) particles are in Half A' or 'two particles are in Half A, two particles are in Half B', that is, by the specification of the two particle numbers N_A (particles in Half A) and N_B (particles in Half B).

It is then easy to see that the macrostate ($N_A = 4$, $N_B = 0$) corresponds to just one microstate, namely ($x_1 = $ Half A, $x_2 = $ Half A, $x_3 = $ Half A, $x_4 = $ Half A). In contrast, the macrostate ($N_A = 2$, $N_B = 2$) corresponds to six microstates, namely ($x_1 = $ Half A, $x_2 = $ Half A, $x_3 = $ Half B, $x_4 = $ Half B); ($x_1 = $ Half A, $x_2 = $ Half B, $x_3 = $ Half A, $x_4 = $ Half B); $\ldots$ ($x_1 = $ Half B, $x_2 = $ Half B, $x_3 = $ Half A, $x_4 = $ Half A). Accordingly, $W(N_A = 4, N_B = 0) = 1$ and $W(N_A = 2, N_B = 2) = 6$. The numbers of possible microstates W for all macrostates of this toy system are shown in Figure 3.

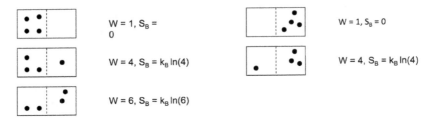

Figure 3 Number of microstates W and Boltzmann Entropy for an ensemble of four particles in box portioned into two halves.

For larger particle numbers and partitions, counting combinations becomes impractical, but we can rely on results from combinatorics to calculate the number of possible ways to distribute N particles into k sets of N_k particles, respectively:

$$S_B \propto \frac{N!}{N_1! \times N_2! \times \ldots \times N_k!}. \tag{3.2}$$

Figure 3.2 illustrates the number of microstates for $N = 100$ particles in a box with $k = 2$. Those examples illustrate that the number of possible microstates W for each macrostate grow rapidly with the particle number. Accordingly, the actual entropy function is based on the more manageable logarithm of W, which also normalises the lowest possible entropy state, corresponding to $W = 1$, to zero. Therefore:

$$S_B \propto \ln\left(\frac{N!}{N_1! \times N_2! \times \ldots \times N_k!}\right). \tag{3.3}$$

Notably, this proportionality relationship preserves additivity, for example, combining two macrostates $S_{B,1}$ and $S_{B,2}$, without loss of accessible microstates, that is, by opening a partition in box with two identical gasses on each side, will simply lead to a combined entropy of $S_{B,1+2} = S_{B,1} + S_{B,2}$. In order to make entropy measure S_B formally compatible with Thermodynamic Entropy (Section 3.1), it needs to be prefaced by a constant with units [J/K]. The final formulation of this statistical entropy is therefore:

Boltzmann Entropy: $S_B = k_B \ln\left(\frac{N!}{N_1! \times N_2! \times \ldots \times N_k!}\right)$,
where $k_B = 1.38 \times 10^{-23}$ J/K.

There already exist various comprehensive analysis of Boltzmann Entropy (e.g., Albert, 2000, Ch. 3; Frigg & Werndl, 2011; Rickles, 2016, Ch. 6). Here, I will only focus on the aspects of Boltzmann Entropy relevant for the two aims of this Element (Section 1): in Section 3.3, I will discuss how Boltzmann Entropy relates to Thermodynamic Entropy (Section 3.1); in Section 3.4, I will discuss how well Boltzmann Entropy performs as a measure of visible disorder.

3.2.2 Gibbs Entropy

The definition of Gibbs Entropy originates in Gibbs (1902/1960). In comparison to Thermodynamic Entropy (Section 3.1) and Boltzmann Entropy (Section 3.2.1), Gibbs Entropy has arguably the most sophisticated conceptual set-up: rather than considering the statistics of different possible microstates of a single system (Boltzmann Entropy, Section 3.2.1), it is based on the consideration of each microstate as a separate system. In other words, the definition of Gibbs entropy is based on the virtual preparation of an ensemble of systems each instantiating a different microstate corresponding to one macrostate. The term 'preparation' here implies that we assume that it is possible (for some entity or just theoretically) to identify all possible microstates (i.e., specifications of phase space vectors) that are compatible with a given macrostate (e.g., the specification of a region of phase space into which all of those vectors will fall) and that an ensemble of systems each instantiating one such microstate is then hypothetically set up.

Without considering a partition, the micro-variable vectors x_i assigned to each particle can assume an infinite and uncountable number of values, therefore this virtual ensemble of systems is also infinite and uncountable. For each macrostate with ensemble Σ, we can then define a density function $\rho_\Sigma(x, t)$, which describes the density of systems within the ensemble whose microstates lie in the infinitesimal volume around x, that is, $\rho_\Sigma(x, t)\ dx$ is the number of systems in the phase space volume $(x, x + dx)$. The probability of finding a system in a given finite volume X of phase space is then:

$$p(X, t) = \int_X \rho_\Sigma(x, t)dx \tag{3.4}$$

The Gibbs Entropy itself is then defined as a conservation function of the probability density $\rho_\Sigma(x, t)$:

Fine-Grained Gibbs Entropy: $S_{G,f} = -k_B \int_X \rho_\Sigma(x, t)\ln \rho_\Sigma(x, t)dx$

Notably, due to the fact that we defined the density function $\rho_\Sigma(x, t)$ over all possible values of x and therefore accepted that the resulting ensemble of systems will be uncountable and infinite, no definition of a partition was necessary.

In order to be able to compare Gibbs and Boltzmann Entropy, we would need to define a coarse-grained version of Gibbs Entropy. This might also be advantageous on independent conceptual grounds, for example, to simplify the integration in (3.4) or if we are interested in phase space volumes that have approximately constant probability density $\rho_\Sigma(x, t)$ through large areas of the allotted phase space. To define a partition-based version of Gibbs Entropy by assuming that density $\rho_\Sigma(x, t)$ is constant within a given phase-space partition w. The corresponding Coarse-Grained Gibbs Entropy is then defined as follows:

Coarse-Grained Gibbs Entropy: $S_{G,c} = -k_B \displaystyle\int_X \rho_{w,\Sigma}(x, t) \ln \rho_{w,\Sigma}(x, t) dx$

For our toy system of four particles in a box of two halves and the phase space vector x_i consisting only of location variables, Figure 5 shows the relevant probability densities and entropies. It's immediately evident that, in this case, the values for the Coarse-Grained Gibbs Entropy $S_{G,c}$ scale like the Boltzmann Entropy $S_B{}^5$. It can also be shown formally that, in the case of equally accessible microstates, that is, an ideal, fully elastic ensemble of particles, Coarse-Grained Gibbs Entropy is formally equivalent to Boltzmann Entropy (Frigg & Werndl, 2011, p. 130). In Section 3.3, I will discuss how Gibbs Entropy relates to Thermodynamic Entropy. In Section 3.4, I will discuss how both statistical entropies perform as a measure for visible disorder.

3.3 Reduction of Thermodynamic Entropy

Within the history of physics, Thermodynamic Entropy (Section 3.1) did not just historically precede statistical entropy definitions (Section 3.2) but also provide the conceptual context for those definitions. In particular, the development of statistical entropy definitions is part of a larger conceptual project of reducing thermodynamics, formulated in terms of the macroscopic (p, V, T)-variables, to statistics physics, formulated primarily in terms of the velocity and position variables, that is, the phase space vector x_i single particles (see above). Callender (1999, p. 359) argues that the complete reduction of thermodynamics to statistical physics is still the 'holy grail' of foundational research in those areas and that, therefore, statistical entropy definitions that further this aim should be preferred over those that do not. While I would not want to set the reduction of Thermodynamic Entropy as the one and only criterion for an acceptable statistical entropy definition, the importance of the thermodynamics-to-statistical-physics reduction project for philosophers and physicists is

---

5 In order to achieve numerical equivalence, one would have to normalise the probability densities for each ensemble accordingly. This is usually done by considering fractional densities over the relevant phase space volumes, for example, to take into account the fact that in the macrostates with one particle in one state and three in the other, only a fraction of the system is in each half.

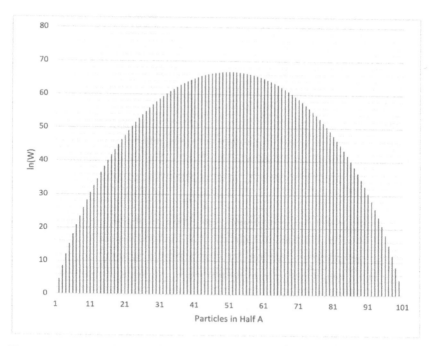

Figure 4 Number of microstates *W* against the numbers of particles Half A, for an ensemble of 100 particles and box partitioned into two halves.

undeniable. Therefore, being able to achieve a reduction of Thermodynamic Entropy is clearly an epistemic advantage for any statistical entropy definition. Furthermore, some entropy-groundings of the Arrow of Time will require an assumption that the constitutive processes of our experience of time are reducible to statistical physics. The question of how well each of the statistical entropies can or could potentially fulfil such an assumption is therefore of direct importance to the derivation of entropy-groundings.

3.3.1 Reductive Potential of Boltzmann Entropy

As Callender (1999) stresses, Boltzmann Entropy is formulated in the terms of statistical particle mechanics and therefore has the potential to straightforwardly reduce Thermodynamic Entropy as long as all the terms in Thermodynamic Entropy have also been given statistical interpretations, for example, as long as temperature *T* is reduced to an expression of the average particle velocity and so on. As such, Boltzmann Entropy straightforwardly slots into the larger project of reducing thermodynamics to statistical mechanics and any difficulties in the reduction are likely to relate to general problems in reformulating thermodynamical variables in terms of statistical mechanical variables.

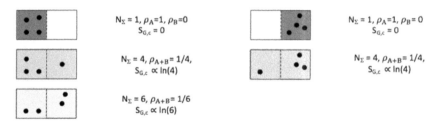

Figure 5 Ensemble sizes, probability densities, and entropy values for four particles in a box partitioned into two halves. The shading represents the probability density in each part of the partition.

However, Lavis (2005) highlights that there is at least one fundamental mismatch between Boltzmann Entropy and Thermodynamic Entropy. In particular, as described in Section 3.1, for a closed system, Thermodynamic Entropy has a single, stable equilibrium value and without intervention from outside, for example, through heat exchange with the environment, will remain at this equilibrium value for all times. In contrast, Boltzmann Entropy has no equilibrium value as such: high-entropy macrostates have overwhelmingly more microstates that can realise them (e.g., Figure 4) and, if all microstates are equally accessible by the system, are therefore overwhelmingly more likely to instantiate. If the condition of equally accessible microstates is true (and this is not unequivocally accepted for 'real' particle systems, as we will discuss in Section 4.3), then it is likely that the system will not deviate from a high entropy value once it is reached. However, this is only an approximation of the strict bi-valued equilibrium notion of Thermodynamic Entropy, and a long development of such a system with equally accessible microstates, one would expect small (and eventually large) deviations from the most likely macrostate to occur.

Given that statistical and deterministic variables are fundamentally different in nature, it seems to me that the approximation of the deterministic equilibrium value of Thermodynamic Entropy through the most likely statistical value of Boltzmann Entropy is not unusual within the wider reduction project (for an extended version of this argument, Sklar, 1993, pp. 348–61). As an epistemic defect, this mismatch is therefore one that is unlikely to inhibit the progress of the wider project of the reduction of thermodynamics, and might well provide an example of the kind of redefinitions one can expect to see within this project.

That said, as we will see in Section 3.3.2, the Coarse-Grained Gibbs Entropy (Section 3.2.2) has a deterministic equilibrium value. However, this comes at the cost of moving away from considering particle systems that directly map onto the ones we think reduce (idealised) thermodynamic systems.

3.3.2 Reductive Potential of Gibbs Entropy

In contrast to Boltzmann Entropy (Section 3.3.1), it is more difficult to argue that either Fine- or Coarse-Grained Gibbs Entropy straightforwardly reduces Thermodynamic Entropy (Section 3.1) or furthers the larger project of reducing thermodynamics to statistical particle mechanics. Firstly, and most glaringly (e.g., Callander, 1999; Frigg & Werndl, 2011), Gibbs Entropy assigns a value of entropy not to macrostates of a single system but to an ensemble of systems all instantiating this macrostate. The object of this definition is therefore quite different from those of the other two entropy definitions and is eminently less compatible with the standard methodology of both thermodynamics (Section 3.1) or other areas of statistical particle mechanics (Section 3.2). In other words, it raises some additional questions about the ontological interpretation of such things as 'virtual ensembles of systems' or a probability distribution over such ensembles. In fact, Frigg & Werndl (p. 128) argue, there are at least two conceptual interpretations available of what Gibbs Entropy actually measures: (i) the abstract probability (frequency) of 'drawing' a system in a given microstate provided a fixed macrostate; or (ii) the average time a system would spend in a particular region of phase space, again provided a fixed macrostate and no energy exchange with the outside world. Independent of the plausibility one assigns to each of these interpretations, they are clearly less intuitively relatable to the particle-mechanics of the system or (as Callander, 1999 argues), to the thermodynamic states of the system. Accordingly, with respect to furthering the larger project of reducing thermodynamics to statistical mechanics, Gibbs Entropy performs less well than Boltzmann Entropy.

However, there is one aspect in which Coarse-Grained Gibbs Entropy straightforwardly matches a feature of Thermodynamic Entropy) that Boltzmann Entropy struggled to recover (Lavis, 2005): it assigns a time-dependent value to $S_{G,c}$, which has a binary equilibrium state, that is, once the function reaches a stable local maximum, it will remain in this state unless an outside intervention takes place. However, this feature only exists for Coarse-Grained Gibbs Entropy; Fine-Grained Gibbs Entropy can be shown to be a constant of motion, that is, it remains the same throughout the system's development. The mathematical proof of this feature involves the application of Liouville's Theorem[6] (Frigg & Werndl, 2011, p. 129) but this fact is also evident from the two possible conceptual interpretations of Fine-Grained Gibbs

---

[6] Liouville's theorem provides a conservation equation for the development of an ensemble of particles subject to a given phase-space density function. Applying it to the density function made available by (3.4) yields the result that Fine-Grained Gibbs Entropy is a constant of motion.

Entropy: if it is interpreted as the probability drawing a specific microstate for a fixed macrostate, then this is independent of the actual instantiation of those states with time; if it is interpreted as the average time a system spends in a given region of phase space, then any time-dependency is eliminated through averaging. There is currently no unequivocal judgement among philosophers and physicists on how those two aspects – the difficulties of finding a conceptual interpretation of Gibbs Entropy and the fact that it recovers a key feature of Thermodynamic Entropy – are to be weighed against each other. In the context of deriving entropy-groundings for the Arrow of Time, which (as we will see in Section 5) often involves significant conceptual reasoning about reduction, Boltzmann Entropy offers more straightforward tools for reductive arguments and has therefore been the statistical entropy most often underpinning such groundings.

3.4 Statistical Entropy and Disorder

As described in Section 1.2, entropy is often conceptualised as a measure for the visible disorder one might find in a child's bedroom or in the shards of a shattered wine glass. For example, Rickles (2016, Chapter 6) displays an actual picture of a child's bedroom with different toys strewn all over the available space. While we might easily be able to conceive of similar paradigmatic examples of visible disorder, for example, a 'well-mixed' pile of Lego bricks or a busy crowd of people in a public square, capturing the notion of visible disorder (and its different degrees) formally has proven more difficult. Firstly, considering the preceding examples, we would (i) like our notion of visible disorder to be anchored by a notion of randomness (Section 2). Secondly, even a closer look at some of the paradigmatic examples reveals that visible disorder is not the only permissible notion of order, and that (ii) we often measure disorder against very specific, scenario-dependent notions of order. In this section, I will first demonstrate aspect (i) and (ii) on the simple, paradigmatic example of a pile of Lego bricks (Section 3.4.1). Then I will discuss how well each of the two statistical entropies performs as a measure for visible disorder (Section 3.4.2). This section will contain a counterexample that shows that statistical entropy is not a good measure for visible disorder. However, I will then argue that it is still possible to view those entropies as measuring deviations from a scenario- dependent state of order, which should be motivated by the physical realities of a given phenomenon (Section 3.4.3). Therefore, statistical entropies can be interpreted as measures of an abstract notion of disorder, but not as measures of visible disorder.

3.4.1 Visible and Abstract Disorder

Let's illustrate aspects (i) and (ii) on the example of a pile of Lego bricks. For this example, the most prominent visible individual properties of the bricks are colour and size, the latter given by the number of interlocking studs rather than a more conventional measure. Additionally, we might want to take into account their position 'in the pile', whereby auxiliary questions about the boundaries of this entity also need to be addressed. Let's assume our Legos are spread out on a square tablecloth of 1 m² size (Figure 6a). We can then link our visibly disordered state definition to a notion of randomness (Section 2) by requiring, say, that a sequence of bricks randomly selected from anywhere on the cloth will result in random sequences of colour and size numbers. Or, if we subdivide the cloth into four subsections of 25 cm² and number those subsections consecutively, the requirement that randomly taking a brick from each subsection in turn results in a sequence of four random colour and size numbers. In this example, it becomes immediately apparent in that there are different ways of linking 'visible disorder' to randomness.

Similarly, there are different ways of defining the contrasting visibly ordered state. 'Ordering' or 'sorting' a pile of Lego bricks can involve different numbers of sub-piles: for example, we might want to make many different sub-piles for each size and colour and put each of those sub-piles in a different place on the tablecloth (Figure 6b). However, both for building and for ordering purposes, it is usually advantageous to keep the number of separate categories smaller than is strictly possible given the set of properties. For example, we might want to group all shades of primary colour together or all 'very small' bricks with less than a given number of studs. Independent of how many different piles we make, the ordered state will no longer be associated with a random sequence in

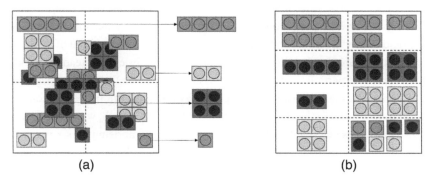

(a)   (b)

Figure 6 a) Selecting a random sequence from a disordered pile of Lego bricks.
b) An ordered ensemble of Lego bricks.

the way that the disordered state was: for example, let's assume we put each sub-pile in one of eight subsections, then we know exactly what sequence of colour and size numbers we obtain by randomly taking a brick from each subsection in turn and the sequence will no longer be random. Therefore, with respect to aspect (i), the multi-realisability with respect to the formalisation of visible disorder through a partitioning of the 'phase space' of visible properties is largely a measure of fine- or coarse-graining of categories and should not affect our ability to decide to formalise visible disorder as a mapping on a random sequence, suitably defined according to a randomness definition as discussed in Section 2.

However, there is a deeper multi-realisability even in this example, which illustrates that visible order/disorder is not the only permissible kind of order/disorder. Legos (and most other construction toys) are usually sold in sets of bricks that all relate to one specific construction project, for example, a pirate ship or a space station, for which building instructions are also provided. Those sets contain a variety of bricks of different colours and sizes, as well as special pieces (e.g., pirate flag or a space helmet) that fall entirely outside the visible property categories we have defined earlier. Another permissible way of ordering a pile of Lego bricks (Figure 6) would therefore be to make piles according to the building project the bricks originally belonged to, for example, a 'pirate ship'- and a 'space station'-pile. Those piles might well be visible disordered, for example, drawing from them would still result in a sequence of randomly sized and coloured bricks, but are ordered according to the non-visible property of 'belonging to set X'. Therefore, with respect to aspect (ii), even in this simple example, we have to recognise that not all notions of disorder match onto visible disorder and that what counts as ordered should be situation dependent, but still justified by the reality of the given situation (in this case, the existence of instructions and pre-compiled sets).

It might seem somewhat facetious to spend that much time discussing what is literally a 'toy example'. However, the typical systems in statistical physics – the field of physics in which the definitions of entropy as a measure of disorder are located – are not that different from piles of Lego bricks: instead of bricks, they consist of particles, and instead of visible properties like colour, size, and positions, we usually consider particle velocity and position. Instead of considering how N different bricks can be distributed on a tablecloth, we will consider how N particles, each with a phase space vector x_i comprising its six position and velocity components, can be distributed through a phase space region with different partitions. Therefore, we would expect that a visibly disordered state in such a system of particles can be shown to be mapping onto a random sequence with respect to those visible properties, much as the disordered state of Lego bricks has been shown to do earlier. In fact, we usually consider gases, with particles strewn through the available spatial space and moving at different velocities to be a paradigm of a visibly disordered state and crystals, with particles fixed in symmetrical positions and all at the same (low or zero) velocity, to be a paradigm of a visibly ordered state of a system.

However, in Section 3.4.2, we will see that particles can have non-visible properties, too, for example, spin, rotational momentum, interactive properties, and so on, and that it seems to be orderings over such abstract phase spaces that should be considered if entropy should be seen as a measure of disorder (Section 3.4.3).

3.4.2 Statistical Entropy as a Measure of Visible Disorder

As discussed in Section 1.1, statistical entropy has been associated with visible disorder from the very moment of its inception. However, it should be noted right away that the kind of systems usually discussed in statistical mechanics, and discussions of entropy in particular, are ones that are best described by the visible properties of velocity (momentum) and position, that is, whose microstates are located in the standard six-dimensional $(\mathbf{x}, \mathbf{p})$-phase space. The particles in those systems are not usually taken to be directly visible to the human eye, of course, but it is assumed that they would seamlessly scale up to visible size without alterations to their dynamics, that is, we could visualise everything that happens in such a system with billiard balls. The fact that it is permissible to only take those two visible properties, namely position and momentum, into account is predicated on the assumption that those are systems of identical (except for those two properties), non-interacting, elastic, point particles. We will see in what follows that even minor violations of these conditions lead to different results with respect to entropy calculations.

However, let us initially consider systems which fulfil the condition that they can permissibly be described in the $(\mathbf{x}, \mathbf{p})$-phase space. For such systems, does Boltzmann Entropy assign maximum entropy states to the states with the highest degree of disorder, which itself is defined through an acceptable definition of randomness (Section 2)? It can be shown (e.g., Frigg & Werndl, 2011, p. 123) that the maximum entropy values of the formula for Boltzmann Entropy correspond to one where $N_1 = N_2 = \ldots = N_k = n$, that is, where the particles are spread out evenly through the partition (this is also illustrated in Figures 3 and 4). Can we translate this state into a sequence comparable to (S1)–(S3) in Section 2? Note that we previously said that we could label the particles from 1 to N. Let's assume we operate a separate lottery box in which we have collected all of the particle numbers and from which we will randomly draw such a number. The probability p_k of drawing a particle which is in a given partition k is then:

$$p_k = \frac{1}{n}. \tag{3.3}$$

We can then write down the partition numbers to produce a sequence that can be analysed through the randomness definitions introduced in Section 2. Assuming that we do not interfere with the particles in the box and always replace the particle number in the lottery box, this drawing process corresponds to a Bernoulli scheme and would therefore be Independent Probability Random (Section 2.1) and Unpredictability Random (Section 2.3)[7]. Given that the probabilities here are equal, we can also be relatively (but not absolutely) sure that this process will produce sequences that are – at least heuristically, if not strictly – Stable Frequencies Random (Section 2.2.1) and Maximum Complexity Random (Section 2.2.2).

Two worries can be raised here: (i) we might be accused of having 'smuggled in' randomness by constructing the sequence through a random drawing from a lottery box; (ii) a random distribution over the partition does not necessarily correspond to a random distribution over other distributions, that is, the multi-realisability of defining categories and piles that we already encountered in the toy example (Section 3.4.1). Worry (i) can be allayed by the fact that the process described earlier robustly maps high-entropy states to random sequences and low-entropy states to non-random sequences, that is, the introduction of the lottery box does not lead to a random sequence for a low-entropy state like the one where all particles are in one box. We can thus be confident that the process of sequence construction preserves the feature conceptualised by Boltzmann Entropy, namely the distribution of particles over the partition.

However, this result leads us back to worry (ii): namely, that randomness and visible disorder relative to a given partition might not correspond to what we would intuitively consider to be visible disorder. Figure 7 illustrates some scenarios in which this would be the case: in the case that the partition is simply too coarse-grained to capture order on smaller scales (Figure 7a) or that the partition is chosen in a way that is at odds with other symmetries of the set-up (Figure 7b). However, questions of coarse- and fine-graining of the $(\mathbf{x}, \mathbf{p})$-phase space might not be as alarming as they first seem. It can be shown that, for typical statistical

---

[7] The fact that the maximum entropy state can be mapped onto maximum unpredictability has led to a separate branch of entropy research in which entropy is described as a lack of information (for review, see, e.g., Frigg & Wendl, 2011, section 3). The information-theoretical formulation of entropy does not directly feed into the derivation of the Arrow of Time and will therefore will not be discussed in detail here.

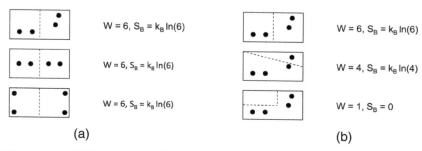

(a)   (b)

Figure 7 a) Different permissible arrangements for the maximum entropy state of a system of two particles in a box portioned into two halves. b) Different partitions leading to different entropy values for the same particle arrangement.

mechanical particle systems with a long development over time, partitions with a similar symmetry converge towards the same high- and low-entropy states (Sklar, 1993, pp. 357–8), that is, barring pathological arrangements, the Boltzmann Entropy for different comparable, but coarser or finer, partitions, will not assign completely different states as low- and high-entropy states.

Secondly, while there might be pathological partitions that would assign high-entropy states to visibly ordered systems or vice versa, those are usually partitions which are recognisable as inappropriate for the given phenomenon. In line with the argument made by Sklar (1993, p. 358), I maintain that there are usually good reasons for why certain kinds of partition appear to be suitable for a certain kind of system. For example, the irregular partitions in Figure 7b are clearly at odds with the symmetry of the relevant phase space region (the box), while a regular partitioning into halves (or quarters or eights) can be justified by this symmetry. Accordingly, for the restricted class of point particle systems considered here, Boltzmann Entropy is a good measure of visible disorder.

For this class of systems whose microstates can be captured in the $(\mathbf{p}, \mathbf{x})$-phase space only, how well does **Gibbs Entropy** work as a measure for visible disorder? It is evident both from the formal definitions of the Fine- and Coarse-Grained Gibbs Entropy, and in particular the ability to define the probability function (3.4), as well as the conceptual interpretations of Gibbs Entropy that we can associate a random sequence with high-entropy states. This can be done by requiring that the probability $p(X_i, t)$ of finding a system in a given phase space region X_i is independent of the probability $p(X_{i+1}, t)$ of finding the system in any other phase space region X_{i+1} (and that those probabilities are approximately equal), thereby fulfilling the Independent Probability Randomness definition (Section 2.1). It can be shown that this is indeed the case for non-interacting, identical particle ensembles (Frigg & Werndl, 2011, 129). A sequence constructed by randomly drawing a system from the ensemble and then writing down their phase pace region or partition will therefore very likely (although not certainly) fulfil the phenomenological randomness definitions, that is, Stable Frequency Randomness (Section 2.2.1) and Maximum Complexity Randomness (Section 2.2.3). We have seen in Section 2.3 that if a system is both dynamical and phenomenologically random, then it is also Unpredictability Random.

Given the discussion of how partitions might distort the mapping between visible disorder and Boltzmann Entropy, the fact that Fine-Grained Gibbs Entropy does not use a partition is often

cited as a major advantage of Gibbs Entropy over Boltzmann Entropy (e.g., Sklar, 1993, p. 355). However, the conceptual difficulty remains that Gibbs Entropy values are actually not assigned to those macrostates but to the whole virtual ensemble corresponding to each macrostate. The conceptual interpretation of what a low-entropy, ordered state means in physical space therefore remains unclear.

While both statistical entropies therefore appear to be good measures of visible disorder for the restricted class of systems that can be described in $(\mathbf{x}, \mathbf{p})$-phase space, that is, systems of identical, non-interacting, elastic point particles, it has been argued that this is no longer the case when even slightly more realistic scenarios are considered (Burgers, 1970; Denbigh, 1989; Frenkel, 1999; Leff, 2007).

In particular, Frenkle (1999, pp. 27–8) presents (among others) a system of rod-like particles, which is otherwise similar to the point-particle systems discussed earlier. However, in addition to the six $(\mathbf{x}, \mathbf{p})$-degrees of freedom, rod-like particles can also rotate around their own axes of symmetry, namely, additional movement on an even smaller, not-immediately visible scale is possible. The correct phase space to describe the dynamics of the rod-like particles is then the $(\mathbf{x}, \mathbf{p}, \mathbf{l})$-phase space, where $\mathbf{l}$ is a vector specifying the rotational momentum in three-dimensional space. If those additional degrees of rotational freedom are taken into account, then it turns out the entropy of the visibly ordered, crystalline arrangement, which allows each particle more 'wiggle room' to rotate around its axes of symmetry, is higher than the entropy of the visibly disordered, liquid phase (Figure 8). Frenkel (1999) demonstrates that there are several such systems with 'invisible' degrees of freedom, for which high-entropy states do not align with visible disorder, and writes (p. 28):

> In fact, we shall see this mechanism returning time-and-again in ordering transitions of hard-core systems: the entropy decreases because the density is no longer uniform in orientation or position, but the entropy increases because the free-volume per particle is larger in the ordered than in the disordered phase.

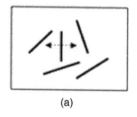 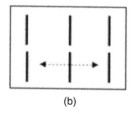

(a)   (b)

Figure 8 a) Low-entropy, visibly disordered macrostate versus b) high-entropy, visibly ordered macrostste of a system of rod-like particles. The horizontal, dashed arrows indicate that there is available 'wiggle room' in each case.

It is notable that the changes we have made to the systems for which statistical entropy did align with visible disorder is relatively minor: we have simply changed the geometry of the particles and thereby added additional degrees of freedom on a scale even smaller than the visible, translation variables. It can therefore be safely assumed that such systems are not rare – in fact, Frenkel (1999) and others (e.g., Georgii & Zagrebnov, 2011; Gobbo et al., 2020) demonstrate that they seem to capture the dynamics of laboratory-prepared systems of molecules with the correct symmetry very well, and that similar effects occur in mixtures of certain types of particles, where the separated, visibly more ordered phase has higher entropy because it allows for more small-scale movement of each individual particle. Accordingly, we can assume that many natural phenomena might well be more similar to the rod-like rather than the point-particle system and that statistical entropy should not be expected to align with visible disorder. As we will see in Sections 4 and 5, this also has consequences for the associated assumption that visible disorder increases with time.

3.4.3 Statistical Entropy as a Measure of Abstract Order

Is it possible to salvage some of the interpretation of entropy as a measure for disorder? I think it is, if we make sure that we eschew any notion of visible disorder and, instead, consider a more abstract notion of order within a phase space that is appropriate to the phenomenon under investigation. We already saw that abstract order can exist even in everyday examples (e.g., in the form of project-based Lego sets), so that this redefinition of the notion of order is not a distortion of the established meaning of the phrase.

It is also important to notice that the phase space that the entropy is computed in is not arbitrary: the degrees of freedom that need to be taken into account are given by the physical realities of a phenomenon, for example, the additional rotational degrees of freedom of the rod-like particles that arise from their specific asymmetries. Similarly, any partition we would like to impose on this system to compute a coarse-grained entropy should be reasonable given the scale and symmetries of the problem. As such, entropy could be interpreted as a notion that measures disorder within an appropriately defined phase space and over an appropriately chosen partition. This definition does not lend itself to the kind of conceptual arguments based on visible order and disorder we discussed in Section 1 but does seem to allow us to move entropy away from being a notion whose interpretation is strictly tied to very simplified point particle systems.

4 Two Versions of the Second Law

In its simplest formulation, the second law of thermodynamics[8] states that, in an isolated system, entropy increases with time. Conversely, the law states that entropy does not spontaneously decrease with time. Here, being isolated implies that there is no heat or other energy transfer in and out of the system. In the following we will make this a standard condition for the laws presented, that is, unless otherwise stated, they only apply to isolated systems. We will use the term 'spontaneous' to refer to any process that happens in an isolated system, that is, without energy transfer in and out of the system.

However, as described in Section 3, there are two fundamentally different kinds of entropy the second law of thermodynamics could refer to: Thermodynamic Entropy (Section 3.1) and Statistical Entropy in the form of either Boltzmann Entropy (Section 3.2.1) or Gibbs Entropy (Section 3.2.2). As Boltzmann Entropy is conceptually easier to interpret (Section 3.3), we will focus our discussion here on this form of Statistical Entropy. However, I will highlight any significant differences that would arise if Gibbs Entropy was chosen instead. Depending on which kind of entropy the second law applies to, its derivation and formalisation are very different: as we will see in what follows, in the case of Thermodynamic Entropy, the empirical second law is derived inductively as an analytical law (Section 4.1); in the case of Boltzmann Entropy, the statistical second law follows deductively from the definition of this entropy and is a statistical law (Section 4.2). Furthermore, the deductive derivation of the statistical second law from the definition of Boltzmann Entropy requires the introduction of an additional assumption about the particle mechanics of the systems under consideration, namely that of equal accessibility of all microstates. The veracity of this Equal Accessibility Assumption will be discussed in Section 4.3. Furthermore, the statistical second law does not

---

[8] The first law of thermodynamics is the requirement of energy conservation, and the third law of thermodynamics fixes the initially entropy value, usually requiring that it is zero at absolute zero temperatures. A zeroth law of thermodynamics is sometimes added, requiring that if thermal equilibrium is an expansive property, namely, if two systems are both in thermal equilibrium with a third, they are also in thermal equilibrium with each other.

predict a monotonic entropy increase but allows statistical deviations from this course. The implications of this feature have featured large in debates on the suitability of the statistical second law as grounding for the Arrow of Time and will be discussed in Section 4.4.

It is notable that the second law is already a dependent notion: as foreshadowed earlier, it can be derived, inductively or deductively, from different kinds of entropy. Accordingly, the discussion from now on will focus on comparing different derivational routes (i.e., as mapped in Figure 1) rather than different concepts. Instead of focusing on properties and interpretations, we will therefore compare assumptions that have to be made during different derivational steps and discuss the implications of different derived concepts. As such, the switch in dialectic between the first two content sections of this Element and the remaining two is evidence of the dependent nature of both the second law and the Arrow of Time.

4.1 The Empirical Second Law of Thermodynamics

The empirical second law of thermodynamics is derived from the observation that certain processes occur spontaneously in nature while their time-reversals only happen if energy is expended on the system, that is, the time-reversals do not happen in isolated systems. Prominent examples used throughout this Element and in many discussions of the second law are the spreading of a gaseous substance through a room or the shattering of a wine glass. Those processes happen spontaneously in one direction (in the case of the wine glass, the natural disintegration process would be on a fairly long timescale, of course), but their time-reversed courses would take an outside intervention into the system, that is, breaking the system's isolation and expending energy on it. They are therefore also called irreversible processes, whereby the term only refers to spontaneous reversal.

Furthermore, if we analyse those empirically irreversible processes in terms of their thermodynamic (p, T, V)-variables, then it becomes apparent that the common processes are associated with increasing or constant Thermodynamic Entropy (Section 3.1), while the forbidden reverse processes are associated with decreasing Thermodynamic Entropy. Empirical data therefore shows that, for closed systems, Thermodynamic Entropy S_{TD} monotonically increases or stays constant:

Empirical Second Law: $\frac{dS_{TD}}{dt} \geq 0.$

It is worthwhile pointing out that the irreversibility captured by the empirical second law only applies to spontaneously occurring processes, namely, to processes that do not receive or lose outside energy through heat or work. It is possible, of course, to repair a broken wine glass or to suck dispersed gaseous molecules back into a small space using a suction pump with an external energy supply. In other words, as stressed at the beginning of the section, the empirical second law only applies to isolated systems.

The derivation of the empirical second law is inductive: we have observed – so far – that thermodynamic entropy always increases in spontaneous processes in isolated systems and, therefore, we expect this to be the case in the future as well. The empirical second law codifies this expectation. It is widely known that induction is not a means of logically valid reasoning and that one can imagine scenarios in which well-established inductive laws might suddenly become invalid (e.g., the Problems of Induction posed by Hempel, 1945 and Goodman, 1955/1983). There can be no logical certainty that an inductively derived law will really be correct for all future instances, that is, while we have never so far observed a process with spontaneously decreasing entropy in a closed system, we have no other reason than that very fact to assume that this will also be true in the future. Due to its inductive derivation, the empirical second law is therefore subject to this fundamental uncertainty inherent in inductive reasoning.

However, while philosophers of science have discussed the validity of inductive reasoning in science (for a prominent sceptical view, see, e.g., Popper, 1959/2002), there is little doubt that many important and fully accepted scientific results are inductive laws. An example of an equally directional law would be the law of gravitational attraction, which posits that the gravitational force is directed towards masses and is proportional to each mass. Notably, the inductively derived directionality here is not one of 'up' and 'down' as we experience it on Earth, as the fact that all objects on Earth seem to fall towards it is a consequence of the magnitude of the mass of the planet compared to that of the objects on it (e.g., Sklar, 1993, p. 389). However, the fact that masses attract each other (rather than repel or remain unaffected) is an inductively derived fact and one which we are willing to view as load bearing, namely, make the basis for further derivations and definitions. As such, the inductive derivation of the empirical second law is epistemically unproblematic, in the sense that it does not raise any problems beyond those inherent in inductive reasoning itself.

The inductive derivation of the empirical second law limits its possible explanatory functions. In particular, it cannot be drawn upon to explain the existence of irreversible processes in nature since it is simply a codification of this very fact. However, it can be drawn upon to explain the specific develop-ment of a given system or process by referring to the necessity of that process obeying this particular law of nature. To return to our earlier comparison example, the law of gravitational attraction similarly cannot explain why masses attract each other, as it is merely an expression of this fact, which we accept on empirical evidence; however, it can explain why a given mass behaves in a certain way in response to the gravitational forces it experiences. This distinction between the explanation of directionality itself and the explanation of directed behaviour might seem pedantic but will become more pertinent when we compare the explanatory function of the empirical second law to that of the statistical second law (Section 4.2).

4.2 The Statistical Second Law of Thermodynamics

How can we derive the statistical second law of thermodynamics from the definition of Boltzmann Entropy (Section 3.2.1)? Let's recall that Boltzmann Entropy is essentially a measure for the number of microstates that can instantiate a given macrostate. Therefore, higher entropy values can be instantiated by more microstates; in fact, as we have seen in Section 3.1, for systems with a large number of particles, by many, many more microstates. Let's make one further assumption about the dynamics of the system we are considering:

Equal Accessibility Assumption: During the time-development of a given system (to which the statistical second law should apply), the system is (at least approximately) equally likely to access any of its possible microstates.

We have called this the Equal Accessibility Assumption; in Section 4.3, we will see that it has been given different names as well. If the Equal Accessibility Assumption is true, then macrostates with high entropy become overwhelmingly more likely to instantiate than macrostates with low entropy values since there are simply so many more microstates that can instantiate them. Accordingly:

Statistical Second Law: $P\left(\frac{dS_B}{dt} \geq 0\right) \gg P\left(\frac{dS_B}{dt} < 0\right)$,

that is, at any point in the development of an isolated system for which the Equal Accessibility Assumption holds, the probability that the entropy is increasing is much higher than that the entropy is decreasing.

The statistical second law is clearly very different from the empirical second law (Section 4.1). Firstly, it is derived deductively from the definition of Boltzmann Entropy and the Equal Accessibility Assumption. As such, it avoids the inherent problems of induction that could be asserted for the empirical second law. However, it introduces an additional assumption in the form of the Equal Accessibility Assumption. Therefore, for a given class of systems, whether the deduction of the statistical second law is valid depends on whether the Equal Accessibility Assumption holds for this class of system. In the context of the derivation of the Arrow of Time, one needs to show that the assumption is valid for the system we are basing this derivation on, in the first instance, systems of many atom-like particles. In Section 4.3, we will see that proving that the Equal Accessibility Assumption is true for such systems has occupied physicists and philosophers of physics since the inception of the statistical second law in Boltzmann (1872).

Secondly, in comparison to the empirical second law, the statistical second law has different implications for the expected behaviour of systems to which it applies. It is important to note here that while the statistical second law itself was derived deductively, the underlying deductive reasoning is a reasoning with probabilities, as is evident in the formalisation of the law as a comparison between probabilities. As such, the law only states what is more likely to happen, rather than what is definitely going to happen. Since it has only statistical validity, deviations from the most likely behaviour of entropy increase can occur; for systems with a long development time, they would even be expected to occur. Therefore, the statistical second law does not predict a monotonic entropy increase, it only states that, at any point in the development of a system to which it applies, an entropy decrease is much less likely to occur than the system remaining at constant high entropy or the entropy increasing. In particular, a prolonged period of overall entropy increase (albeit with statistically expected dips) would only be predicted by the statistical second law if the system starts in a low-entropy state. We will discuss this conditionality on the system's initial conditions in more detail in Section 5. The potential ramifications of deviations from a monotonic increase or constantly high state of entropy have been discussed extensively. However, as we will see in Section 4.4, the purely theoretical implications are not as grave as expected in the last century. Nevertheless, the possibility of such deviations can have a larger influence on whether and how the Arrow of Time should be grounded in statistical entropy (Section 5.2).

As we have seen in Section 3.4, statistical entropy is only a good measure for visible disorder for the restricted class of systems that can be described in the $(\mathbf{x}, \mathbf{p})$-phase space, namely, identical, non-interacting, elastic, point particles. For those systems, the statistical second law also predicts that visibly disordered macrostates have a higher probability of instantiating, for example, we should expect to see a development towards visible disorder in such systems (if they are isolated). However, as pointed out in Section 3.4.2, even small deviations from the conditions put on those systems (e.g., moving from point-like to rod-like particles) means that statistical entropy is no longer a good measure for visible disorder. As such, the statistical second law should not be seen as predicting that visibly disordered states are more likely to instantiate in nature.

It is possible to derive a statistical second law for Coarse-Grained Gibbs Entropy (e.g., Sklar, 1993, pp. 199–207). However, like Gibbs Entropy itself (Section 3.2.2), the law then refers to the behaviour of virtual ensembles of systems, which makes it less suitable as a grounding for the Arrow of Time. However, it is worth noting that the assumptions made during this derivation and the implications of the statistical nature of the law are the same as the ones arising during the derivation of the statistical second law from Boltzmann Entropy.

4.3 The Equal Accessibility Assumption

In discussion of the statistical second law, from the beginning, it has been noted (for an analysis of this early debate, see, e.g., Brown *et al.*, 2009) that the mechanics of the particles that make up a typical statistical mechanics ensemble (e.g., elastic collisions between point-particles of a given initial velocity) are symmetric and not themselves irreversible. That is, in contrast to the spreading gas or the shattering wine glass, there is nothing to indicate that the time-reversals of such particle collisions do not or cannot happen spontaneously. In other words, if we viewed a recording of a given particle collation backwards, we could not tell that this was a backwards recording and not a 'normal' recording of the collision with reversed initial velocities. Therefore, from the mechanics of the particles alone, we would not expect an asymmetric probability relation like the statistical second law to arise.

The statistical asymmetry found in the statistical second law must therefore arise from the statistics of the interaction of particles within the ensemble rather than from considerations of individual trajectories and collisions. Statistics of interactions here refers to the likelihood of a particle experiencing a collision with a certain outcome, for example, reversing its velocity vector; speeding up, or slowing down. In particular, those interaction-statistics need to make true the Equal Accessibility Assumption, which was crucial in the derivation of the statistical second law (Section 4.2), and in doing so, introduce the asymmetry we see in the statistical second law. There are therefore two separate questions to address when assessing the Equal Accessibility Assumption: (i) whether it is possible to find theoretical interaction-statistics that render all microstates equally accessible (Section 4.3.1) and (ii) whether such interactions are likely to occur spontaneously in 'real', or at least defensibly 'idealised', ensembles of particles (Section 4.3.2). In Section 4.3.2, we will also discuss whether the interaction-statistics put forward in response to (i) introduce an artificial asymmetry.

4.3.1 The Stosszahlansatz

As an answer to question (i), Boltzmann (1972) proposed the *Stosszahlansatz* as a suitable assumption about the interaction-statistics of particles in a given ensemble to which the statistical second law should apply. The *Stosszahlansatz* requires that for each particle in the ensemble we define a region S which it will traverse in a given time t; the size of S will depend on the particle's velocity v_i, of course. We then make the crucial assumption that the likelihood of this particle experiencing a collision with a second particle with velocity v_k during time t is directly proportional to the number of particles with velocity v_k in S. The same is true for all other velocities of particles in the ensemble. In other words, the *Stosszahlansatz* assumes that each particle is equally likely to collide with any other particle in its vicinity, independent of previous inter-actions or the relationship of their respective velocities. Notably, while this might sound like a symmetric set-up, it does introduce a time-asymmetry in that previous collisions do not affect future collisions in a realistic way, for example, the fact that two particles might just have collided with each other and are now unlikely to do so again, is discarded. In Section 4.3.2, I will return to this point.

A conceptual argument can be made that such interaction-statistics will indeed make all microstates equally accessible, that is, make true the Equal Accessibility Assumption. Imagine that we have again set up a number of particles in a box of two halves, Half A and Half B (Figure 9). As before (e.g., Figure 3), a microstate is then an assignment of location values of Half A or Half B to each particle. In order for each microstate to be equally accessible, it needs to be the case that each collision makes it equally likely for a particle to be in Box 1 or Box 2 after the collision. This will be the case if: i) that half of the velocity directions a particle can have after a collision will lead to the particle remaining in the half of the box it is in and half will lead to it moving into the other one; ii) that the velocities are initially equally distributed, namely, that half of the particles have velocities that, in an elastic collision, would send the other particle careering off into the opposite half and half have velocities that would result in the particle staying in the same half of the box; iii) as decreed by the *Stosszahlansatz*, each particle is equally likely to collide with any particle in its vicinity, which is roughly equal to the half of the box it is in, and therefore equally likely to be jolted into one or the other half during a collision. Those assumptions mean that each particle has equal probability of being in each box after each collision and that therefore each microstate, which is a combination of those values, is equally accessible, as required by the Equal Accessibility Assumption.

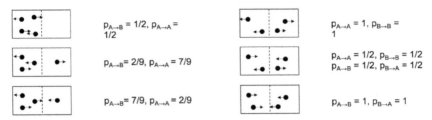

Figure 9 *Stosszahlansatz*-dynamics in a box with halves A and B and four particles. The arrows indicate the direction a particle will jolt (*'stoss'*) its next collision partner. Assuming no interactions with the box's boundaries and that particles only change halves during collisions, the figure shows the probabilities for the next collision in each half resulting in a particle leaving this half or staying in it. It is easily apparent that collisions that lead to macrostates with equal or higher entropy are always more likely.

It is then also easy to conceptualise how such *Stosszahlansatz*-interactions make the transition from a low to a high Boltzmann Entropy state much more likely than the other way around. Let's assume the particles are initially all in Half A (Figure 9). Each particle then has a very high probability of experiencing a collision, given that there are many particles in its vicinity and, according to the *Stosszahlansatz*, that means that a collision is likely. Given the equal distribution of velocities prescribed earlier, after the first round of such collision events, about half of the particles will therefore have been sent into Half B. Under this new situation, particles in both halves of the box will experience collisions but the numbers of potential collision partners in their vicinity are lower (Figure 9). Similarly, it is always less likely that a particle will change back into a half with currently more particles in it than the other way around. Therefore, each collision event will tend to move the system from a lower-entropy state to a higher-entropy state. In addition, any minor imbalances between the particle numbers will tend to be balanced out in the next round of collisions, as particles in the half with momentarily more particles in it will experience more collisions than those in the half with momentarily fewer particles in it, until the number have balanced again. Therefore, the high-entropy state is not only the one towards which the system will tend over time but also a stable state, in the sense that the system will tend to correct back to it.

Such simplified models and similar thought experiments (e.g., Ehrenfest & Ehrenfest, 1907; 1909) show that *Stosszahlansatz*-like assumptions can be successful in ensuring the Equal Accessibility Assumption is true and, there-fore, guarantee the validity of the statistical second law in a variety of scenarios. Hence, we can answer in response to question (i) that it is possible to find theoretical interaction-dynamics that make true the Equal Accessibility Assumption. However, this does not imply that such interaction dynamics also exist in real, or acceptably idealised, ensembles of particles, that is, it does not imply a positive answer to question (ii).

4.3.2 Objections to the Stosszahlansatz

Let's first consider an ensemble of elastic, identical particles, namely, the ideal gas that is the paradigm scenario in statistical mechanics. Even for this idealised scenario, the *Stosszahlansatz*-assumption that each particle is equally likely to collide with each other particle in its vicinity, and that the probability for a collision only depends on the local particle density, is clearly unrealistic. Even for completely elastic collisions, the velocities of two particles that have collided with each other will become correlated and the probability of those two particles colliding again will become dependent on the precise modalities of that collision. For example, consider two particles colliding at relatively slow speeds. By the law of momentum conservation, after the collision, the particles will be on reversed trajectories, moving slowly away from each other. Despite the fact that the two slow-moving particles will still be in the vicinity of each other for some time to come, this previous collision means that they are currently moving in opposite directions, namely, their velocities have become anti-correlated, and they will not collide again (unless deflected in some other way). However, the assumptions of the *Stosszahlansatz* mean that this cannot be taken into account and that the probability of those particles experiencing another collision is computed as if the previous collision never happened. This could be rephrased to state that the *Stosszahlansatz* requires a deliberate discarding of any information about potential velocity correlations between the particles in the ensemble (Brown *et al.*, 2009). Given that velocity correlations are clearly a physical feature of particle collisions, this raises the worries that the *Stosszahlansatz* i) introduces an artefact into the situation; and ii) that it is precisely this artefact that leads to the asymmetry that is captured by the statistical second law.

In particular, by discarding information about past collisions, the *Stosszahlansatz* introduces a time-asymmetry, which – as discussed earlier – is not present in the mechanics governing the particle collisions. One could therefore argue that the *Stosszahlansatz* artificially distorts the system towards the time-asymmetry found in the statistical second law. This concern was first raised by Loschmidt (1876) and is therefore often titled Loschmidt's objection.

How damaging is Loschmidt's objection? It is undoubtedly true that velocity-correlations are a feature of all particles' microdynamics that the *Stosszahlansatz* deliberately ignores. However, it can be argued that these correlations become less important the larger the particle ensemble itself is and the longer we observe it. In particular, in an ensemble of many particles and many collisions, the trajectories of any given particle will quickly become determined by other factors rather than that of any given past collision. Hence, after a sufficiently long time-period, it is a reasonable approximation to treat any two particles as uncorrelated again. The length of time-period will depend on the size of the ensemble and the frequency of the collisions within it: the larger the number of particles and the larger the number of collisions, the shorter the time for which velocity correlations are expected to last. Figure 9 illustrates this argument.[9]

Accordingly, Loschmidt's objection can be interpreted as a condition on the kind of systems to which the statistical second law is likely to apply: namely, those with many particles, frequent particle interaction, and a long duration. Furthermore, the stronger any correlations between the particles of a given ensemble are, the more of a falsification of the interaction-statistics the *Stosszahlansatz* becomes, and the more doubtful we should be that the Equal Accessibility Assumption and, therefore, the statistical second law, obtains. In particular, considering the example of rod-like particles in Section 3.4.2, the Equal Accessibility Assumption would mean ensuring that both the $(\mathbf{x}, \mathbf{p})$-degrees of freedom as well as the small-scale $\mathbf{l}$-degrees of freedom are equally accessible. Therefore, one would have to ensure that particles did not just collide with each other in a certain way but also that they are simultaneously able to rotate freely around their symmetry axes. As Frenkel (1999) recognises, in such situations, it is often not possible to ensure all states are equally accessible and the dynamics of the system will be dominated by trade-offs between the different degrees of freedom. This is less of a problem in the context of statistical physics proper, where large ensembles of fully elastic particles are the typical subject of study, but will become more relevant in the context of asking whether the statistical second law applies to more complex objects, including the universe as a whole (Section 5.2).

4.4 Statistical Deviations from Entropy Increase

As derived in Sections 4.2 and 4.3, for systems with large numbers of particles and frequent particle interactions, macrostates with statistical entropy values become overwhelmingly more likely than those with lower statistical entropy values. This is evident even for a small system of 100 particles (Figure 4), whose maximum-entropy macrostate is more than 10^{28} times more likely to instantiate than its minimum-entropy macrostate. For more realistic systems

---

[9] Brown *et al.* (2009, p. 188) present a formal version of this argument.

with particle number on the magnitude of 10^{23} or more, the probability differences will be even larger. A transition from a high statistical entropy to a low statistical entropy state is therefore a statistically highly unlikely exception.

That said, in a system governed by probabilistic laws, even unlikely events are expected to occur, if a sequence of enough such events instantiates. In other words, if we observe a system of particles long enough, then even low-entropy states should eventually occur with a finite probability. For the 100-particle system, it would take about 10^{30} 'tries' for one of the lowest-entropy states to instantiate with more than 90 per cent likelihood.[10] This back-of-the-envelope calculation can be formalised to prove that any closed dynamical system (e.g., like the simplistic particle systems most of the discussion is based on) will eventually return infinitely close to its initial state. This result was first published in Poincaré (1890) and is known as Poincaré's Recurrence Theorem.

To get an idea of the timescales involved here, let's assume that each 'instantiation' of a microstate takes about 1 s to complete. To have a 90 per cent likelihood of a low-entropy state occurring, the system with 100 particles would have to evolve for about 10^{22} years, namely, longer than the current estimated age of the universe. It is worthwhile emphasising what this means for our expectation about the world: in this universe, we should not expect (i.e., assign a likelihood of more than 90 per cent) to see 100 particles spontaneously returning to one half of a box, much less to see typical systems of many more atomic particles spontaneously assuming a low-entropy state.

As such, the statistical second law leads to expectations that tally very well with the observations that have been used to inductively derive the empirical second law (Section 4.1). However, the fact that the probability of such an event occurring is extremely small still leaves open the possibility of it occurring against overwhelming odds. In other words, the statistical second law is not exceptionless, while the thermodynamic second law is exceptionless. Accordingly, we re-encounter the problem discussed in Section 3.3: the statistical second law and the thermodynamic second law are qualitatively different and any proposed reduction of the latter to the former will have to bridge this difference. In the context of this Element and our Aim 2 of analysing different routes of derivation of entropy-groundings, it is important to note that this conceptual difficulty is not resolved along those routes. As discussed in Section 3.3, the overall reduction project remains unfinished, and entropy-reduction is one of the unfinished key issues.

---

[10] Using $P(X, n) = nP(X)$ for the probability of an event X with probability $P(X)$ occurring in a sequence of n independent trials.

4.4.1 Maxwell's Demon

The possibility for deviations from the most probable course of a system as predicted by the statistical second law leaves open the possibility that one may find a way of amplifying this extremely small (but finite) probability to a larger one by a series of energy-preserving manipulations on the system. The possibility of exploiting statistical deviations by amplifying them within a closed system, namely, without expending any externally generated energy on the system, underpins a sequence of influential thought experiments proposed throughout the twentieth century: namely, various versions of Maxwell's Demon. In recent years, Earman and Norton (1998, 1999) and Norton (2013) have criticised the ongoing importance of Maxwell's Demon in the literature on the foundations of statistical mechanics as overly focused on the extraneous details of this thought experiment, rather than on actual foundational problems. I agree with those authors that the highly hypothetical nature of Maxwell's Demon thought experiments renders them of limited usefulness to the investigation of what the statistical second law actually entails. Accordingly, I will keep the discussion here relatively brief and view those thought experiments as means of highlighting the existence of the (extremely small) probability of low-entropy transitions by presenting imaginative, hypothetical scenarios in which such an instantiation could be amplified to have larger consequences, rather than considering those scenarios as yielding insights into the foundations of statistical physics.

In Maxwell's (1867) original thought experiment, we start out with the following scenario: a bi-partitioned box of fast and slow particles in a maximum entropy macrostate, that is, with equal numbers of fast and slow particles in each half. One of the minimum-entropy macrostates of this system is the one where all fast particles are in one half and all slow particles are in the other. Accordingly, the transition from an even distribution of particles with different velocities throughout the box to one where slow and fast particles are separated into one half each is extremely unlikely to occur spontaneously. Note that this does not imply that it would be impossible to separate the particles if we allow work to be done on the system, for example, by installing a selectively permeable membrane or by heating one half and cooling the other. However, this would require outside interventions, which would themselves consume energy. The system would therefore not be closed anymore, and we would not expect any version of the second law (Sections 4.1 and 4.2) to apply.

Maxwell's (1867) thought experiment hinges on the introduction of an entity that could manipulate the system but not consume any energy in doing so, namely, a demon. Maxwell's Demon is able to track all of the particles in the box perfectly and to operate a frictionless partition between the two halves without energy expenditure. Therefore, Maxwell's Demon is extraordinary in two ways: like Laplace's Demon (Laplace, 1826/1951), it has the ability to perfectly observe all the physically relevant components of its world (i.e., the particles in the box); additionally, it also possesses a metabolism that does not consume any energy when exerting itself (i.e., by lifting the partition). Maxwell's Demon then manipulates the particles in the box in the following way: whenever a slow particle moves towards Half A, it will quickly lift the partition and let it through, and whenever a fast particle approaches Half B, it will do the same. At all other times, it will leave the partition lowered. Given that we expect some statistical fluctuations between the particle numbers in the halves (as described earlier), over time, the demon will be able to trap all slow particles on one side and all fast particles on the other. Accordingly, by virtue of its superior knowledge of the particles' trajectories, it will be able to force a transition from a high- to a low-entropy state. In addition, by virtue of its superior metabolism, it will also be able to do this without expending energy. Given that it will now be possible to exploit the energy differential between the particles in the two halves of the box to, say, drive a little flywheel, Maxwell's Demon has violated the first law of thermodynamics, that is, it has violated energy conservation!

The original thought experiment seemed to have been designed to show that the notion of statistical entropy and the statistical second law are grounded in the epistemic restrictions inherent to the human mind. Conversely, a demon as described earlier would not face such restrictions and would have no need for a notion like entropy. However, more recent discussions of Maxwell's Demon have usually focussed on a different aspect: namely, its ability to use statistical deviations from a monotonic entropy increase – as predicted by the statistical second law – to engineer a violation of energy conservation. However, in order to show that the statistical second law would allow violations of the law of energy conservation, one would have to show that this is not just the case for a demon with an already energy non-conserving metabolism. As such, many later versions of the thought experiment have focused on replacing the demon with more realistic mechanical mechanisms. A number of such mechanisms have been put forward and are comprehensively surveyed and illustrated by Earman and Norton (1998, 1999): for example, several thermo-mechanical machines introduced by Smoluchowski (1912, 1914), including a particle-sized trapdoor that opens in one direction only; a particle-sized paddle that is ratcheted to turn in one direction only; and a mechanical shuttle that transfers heat excess between different reservoirs.

It is important to note that all those machines work on the assumption that unlikely, but expected, fluctuation from a statistically expected distribution of microstates can be exploited to engineer violations of energy conservation and to amplify initially small transitions from high- to low-entropy states into larger deviations. However, given that – under both interpretations – the second law of thermodynamics only applies if there is no energy expenditure on the system, it is crucial that the machines presented here be viewed as themselves operating without energy expenditure. However, in each case, convincing arguments can be made against this assumption: for example, in case (i), the trapdoor must operate on a spring that stores energy; in cases (i) and (ii), energy needs to be expended to maintain the particle-sized mechanisms against thermal fluctuations themselves; in case (iii), energy needs to be expended to monitor the temperature fluctuations in the reservoirs. With respect to the latter case, Szilard (1929/1972) demonstrated that any measurement of microstate fluctuations can be seen as itself increasing entropy and using energy. Accordingly, I agree with Earman and Norton (1999, p. 5), that:

> [O]ne tentatively assumes that the Second Law is secured from Demons and then one deduces what the hidden entropy cost of demonic operation must be. If one can find an independent justification for this cost, one then posits it independently and infers back to the protection of the Second Law. It is our perception that this research programme has been a disappointment if not an outright failure.

While the devising of mechanical demons resulted in some intellectually interesting devices, those attempts seem to have added little to the conclusions we can securely draw from the initial thought experiment: namely, that an entity that precisely monitors the microstates of all particles involved has no need to describe the system in terms of macrostates and their associated entropies. However, that does not imply that the microdynamics Maxwell's Demon would observe would not be compatible with the statistical second law. Nonetheless, the fact that statistical deviations from a monotonic increase of entropy are expected under the statistical second law will have consequences for the suitability of this version of the second law of thermodynamics as a grounding for the Arrow of Time, and we will return to this feature in Section 5.

As in the comparison of different entropies (Section 3), the suitability of a given version of the second law of thermodynamics for a given task needs to be assessed against the epistemic demands of this task. In Section 5, we will use the comparison of the two versions as presented earlier to assess their ability to serve as a grounding (Section 1) for the Arrow of Time. This does not imply that both versions cannot be assessed differently in different contexts: for example, clearly, the statistical law of thermodynamics is the more fundamental law in the larger project of reducing thermodynamics to statistical mechanics (e.g., Robertson, 2020); it could be argued that both versions are contingent on a 'Minus First Law' codifying the requirement for certain systems to approach an equilibrium state (Brown & Uffink, 2001). However, the characteristics highlighted in my analysis will be the ones that are most relevant to the evaluation of each version's suitability as a grounding for the Arrow of Time.

5 Three Arrows of Time

Based on human experience, it seems to be undisputed that the passing of time is an asymmetric phenomenon (for review, see, e.g., Le Poidevin & McBeath, 1993; Albert, 2000, chapter 6): we age forward, but not backward; we have memories of the past, but not of the future; and we can causally influence future, but not past, events. In a less anthropocentric context, our understanding of causality itself also supports the notion that time has a direction: barring pathological cases like strange quantum effects, there appears to be a time-ordering between causes, which need to happen first, and effects, which need to happen after. Conversely put, the directionality of time means that there is (on the level of commonly accessible experience) no such thing as retro-causality. It is also notable that this directionality of time appears to be exceptional if compared to other fundamental notions. For example, (local) space appears symmetric to us both experientially as well as causally.

While it is generally accepted there is an asymmetry to time and that it is evidenced in the experiential and causal time-ordering of events as described earlier, it is more difficult to quantitatively capture and formally define this direction of time. One category of philosophical approaches focuses on the possibility of defining a fundamental way of time-ordering events (e.g., prominently, McTaggart, 1908; Prior, 1967, Mellor, 1998). A second category of approaches focuses on 'grounding' the direction of time in an observable, asymmetric physical property.

As discussed in Section 1, the notion of grounding can be difficult to unpack: while there is some agreement that it establishes a relation where the grounding entity is fundamental to the existence of the grounded one, that is, that the grounded entity is as it is in virtue of the grounding one, there is no unequivocally accepted analysis of what exactly the epistemic properties of an 'in virtue of'-relationship are (for review, see, e.g., Correla & Schneider, 2012). For this Element, we will use a particular unpacking, which seems to capture a) what philosophers and physicists mean when they talk of entropy-groundings of the Arrow of Time and b) does not distort the general notion of grounding as laid out earlier. Namely, following a distinction made by Sklar (1993, pp. 388–96), we will maintain that there are two epistemic components to a physical **grounding**: (i) a **definitional component**, requiring that the behaviour of the physical quantity time is 'grounded in' adequately defines the direction of time as we experience it; and (ii) an **explanatory component**, stating that the behaviour of the physical quantity time is 'grounded in' adequately explains our experience of time as being asymmetric, namely, that it explains the past–future distinction in phenomena like aging, memory, and causality. Thereby, the definitional component will be unpacked as enabling us to assign a time-ordering to a given set of events, that is, to decide which events are 'past' and which are 'future'.

If we are looking for candidates of physical quantities and laws that could ground the asymmetry of time, then entropy (Section 3) and the second law (Section 4) are obvious candidates to do so[11]. However, given that there are different definitions of entropy (Section 2) and different versions of the second law of thermodynamics (Section 3), there are also different routes to deriving entropy-groundings to the Arrow of Time. Any entropy-grounding of the Arrow of Time is therefore a dependent quantity: it depends on the specific definition of the quantities the Arrow of Time will be grounded in. Furthermore, as we will see in what follows, it depends on the kind of systems which are seen as foundational in entropy-grounding the Arrow of Time. As we will see later, the Arrow of Time can be grounded in the entropy of the universe or in the entropy of local processes. Therefore, we can actually identify three Arrows of Time: an Empirical Arrow of Time (Section 5.1) grounded in the empirical second law (Section 4.1); a Universal Statistical Arrow of Time (Section 5.2) grounded in the statistical second law (Section 4.2) applied to the universe as a whole; and a Local Statistical Arrow of Time (Section 5.3) grounded in the statistical second law applied to the local constituting-processes of our experience of time. Note that we will refer to those as Arrows of Time for the sake of brevity but will actually in each case be talking about a specific entropy-grounding of this Arrow.

As in Section 4, our strategy will be to identify the assumptions that go into derivation of each entropy-grounding. Furthermore, using the distinction between the definitional and explanatory components of a grounding introduced earlier, we will identify the different epistemic functions each grounding fulfils. It will become apparent that the Empirical Arrow of Time fulfils the definitional component but not the explanatory one, while the two Statistical Arrows of

---

[11] There are other candidates which we will not discuss in this Element (e.g., section 1; Roberts 2022).

Time fulfil the explanatory component – at least in the sense of having high explanatory potential – but not the definitional one.

5.1 The Thermodynamic Arrow of Time

As we have seen in Section 4.1, spontaneous processes in closed, thermodynamic systems have been observed to only occur in a way that increases the Thermodynamic Entropy (Section 3.1). Processes that would decrease thermo-dynamic entropy, including the reverse of entropy-increasing processes, do not occur spontaneously. From those observations we inductively derived the empirical second law. It is now possible to use the empirical second law as a means to make visible to us the direction of time by linking it to our observations about entropy, namely, by assuming that the direction of thermodynamic entropy defines the direction of time, we have gained a means of deciding which direction the Arrow of Time points at any given moment. Sklar (1993, p. 388) illustrates this process in the following way[12]:

> [Consider] watching a film of a physical process. How can one tell if the film is being run in the proper direction – that is, with earlier projected images being the images of earlier events – and not in the improper, reversed direction that is, with the initial images being of the final events? Unless entropic features play a role, it is argued, one simply cannot tell. ... But if there are entropic features of the processes involved, then one can easily tell if the film is being run in the proper direction.

For example, we know from the empirical second law that a gas will only ever spontaneously increase its volume, that is, in a closed container it will spread through the whole container rather than gather in a corner. If we see a film of a gas spontaneously gathering in a corner of a closed container, we know that we are watching a recording of the real events backwards.

Empirical Arrow of Time: The Empirical Arrow of Time points in the direction of the empirical second law of thermodynamics, that is, we can use processes governed by the empirical second law define the direction of time by time-ordering events.

As the empirical second law governs the behaviour of Thermodynamic Entropy (Section 3.1), we have therefore – on the level of quantities rather than processes – derived an entropy-grounding of the direction of time.

---

[12] This specific quotation is referring to what we will label 'Statistical Arrows of Time'. However, the process of assigning time stamps according to the entropy of a process at a given point can be done using any kind of entropy.

A more pertinent concern might be that the time-ordering defined through the Empirical Arrow of Time could conflict with the time-ordering of processes we consider paradigmatic for our experience of time (e.g., aging, memory-formation, causality). However, observationally, this does not seem to be the case: in the example presented earlier, while we observe the gas spreading, we also grow older; we form new memories of events where the gas was less spread out but do not have any of when the gas will be further dispersed; and we can reliably associate causes with more localised stages of the gas and effects with more dispersed ones. Therefore, the Empirical Arrow of Time fulfils the **definitional component** of a grounding in the empirical second law: it allows us to time-order events by assigning labels like 'past' and 'future' based on a comparison of the Thermodynamic Entropy development linking those events.

There are different ways of interpreting the fact that the directionality of the constituting-processes of our experience of time is the same as the directionality of the empirical second law. On the one hand, we can simply assume that this association between time-experience and thermodynamic processes is another inductively proven fact of the world. As the empirical second law itself, it seems to be a strong inductive association, on par in validity with other inductively derived laws and groundings in the world, for example, the one between the law of gravitational attraction and the up- and down-distinction (Section 4.1; Sklar, 1993, p. 389).

On the other hand, this alignment of the paradigmatic processes of our experience of time could be deductively explained by the fact that those processes themselves are processes for which the Thermodynamic Entropy (Section 3.1) should increase. This would mean that the empirical second law does not just give us a means of defining the direction of time but actually explain our experience of directionality, that is, it explains that our experience of the asymmetry of time results from the fact the processes that make up our experience of time are themselves governed by the empirical second law and therefore have the same asymmetry as the law itself. However, Thermodynamic Entropy and the empirical second law by themselves do not offer much promise of formulating this explanation formally and stringently. It seems unlikely that we would be able to describe processes like aging, causality, and memory formation in terms of the thermodynamical (p, V, T)-space with enough detail to then prove that the Thermodynamic Entropy increases for those processes according to the empirical second law. In and of itself, the empirical second law therefore can be used to define the Empirical Arrow of Time, but not to explain it, that is, it does not fulfil the **explanatory component** of a grounding in the empirical second law, as defined earlier. The fact that we have inductive proof that this definition is consistent with our experience of time is a precondition for our acceptance of this definition, but it is epistemically independent of the definition itself.

However, as we will see in Section 5.2, a grounding of the Arrow of Time through Statistical Entropy and the statistical second law has more potential for offering an explanation rather than a mere definition. Furthermore, we have seen that the different kinds of entropy and the two versions of the second law of thermodynamics are not entirely independent of each other: there is a long-standing commitment among philosophers and physicists to eventually achieve a reduction of thermodynamic quantities to statistical mechanical ones (Section 3.3). That said: as long as this project has not come to fruition, we are left with several different routes to deriving the Arrow of Time and, I maintain, the explanatory potential of a grounding in the statistical second law cannot be counted as an epistemic advantage of the grounding in the empirical second law, namely, of the Empirical Arrow of Time. However, those thoughts highlight the assumption that underpins the thermodynamics-to-statistical-mechanics reduction project: namely, that statistical mechanics is more fundamental than thermodynamics. This implies that an entropy-grounding in Thermodynamic Entropy (e.g., the Empirical Arrow of Time) is less fundamental than an entropy-grounding in Statistical Entropy (e.g., the Statistical Arrows of Time we will derive in Sections 5.2 and 5.3). In fact, should the reduction project come to fruition, then the Empirical Arrow of Time would become redundant. This possibility should itself be seen as a property of the Empirical Arrow of Time, which can be weighed up against its other advantages and disadvantages. At the current moment, however, the reduction of the Empirical Arrow of Time remains a possibility rather than a certainty, and this eventuality should not be given more epistemic weight than any other properties laid out in this section.

5.2 The Universal Statistical Arrow of Time

In Section 4.2, we have demonstrated that the statistical second law can be derived deductively from the definitions of Statistical Entropy (Section 3.2) and the Equal Accessibility Assumption (Section 3.3). The statistical second law states that it is always more likely – for a system that fulfils the conditions of the Equal Accessibility Assumption – that the Statistical Entropy of this system increases or remains at constant entropy rather than that it decreases. As such, we should (usually with a high likelihood) expect that entropy increases or remains at constant entropy with time. In Section 3.3, we have seen that the Equal Accessibility Assumption is not trivially fulfilled by most natural, or even idealised systems. However, there are certain conditions, for example, the ones codified in the *Stosszahlansatz* under which it can be seen as approximately true. Given that there is a general assumption that virtually all processes in the universe can be reduced to the interaction of atomic particles, the statistical second law allows us to define an entropy-grounding based on the entropy of the universe (viewed as an extraordinarily large system of particles):

Universal Statistical Arrow of Time: The **Universal Statistical Arrow of Time** points in the direction of the statistical second law applied to the Statistical Entropy of the Universe, that is, we can use the statistical second law to define the direction of time and to explain our experience of it.

We have therefore derived a grounding of the Arrow of Time in Statistical Entropy.

In complete analogy to the illustration of how the Empirical Arrow of Time (Section 5.1) can be used to define the direction of time, we can use the grounding presented earlier to assign time-orderings to states of the universe if we know their entropy values: a low-entropy state of the universe is likely to have occurred before a high-entropy state. Given the statistical nature of the Universal Statistical Arrow of Time, such assignments will only ever be probabilistic, that is, it is theoretically possible – if unlikely – that a low-entropy state follows a high entropy state (Section 4.4). We will discuss in what follows the consequences of the fact that such time-assignments are non-deterministic.

However, as we will soon see, the assumption that the universe is subject to the statistical second law and that a general entropy increase translates into entropy-increasing subprocesses is not straightforwardly found to be true. In particular, any argument towards its veracity will require two further assumptions. The first assumption is that the universe (i) is the kind of system for which Boltzmann Entropy can be computed and (ii) that it is reasonably similar to the closed systems of identical, elastic frequently interacting particles to which the statistical second law is applicable. This can be formalised as:

Cosmic Entropy Assumption: It is possible to compute the Statistical Entropy of the universe, and its entropy development will be subject to the statistical second law.

There is currently no consensus view on whether the Cosmic Entropy Hypothesis applies, nor are there any unequivocally accepted calculations or estimates of the universe's Statistical Entropy. On the face of it, the universe does not appear very similar to the systems of interacting, closed, identical particles for which the Boltzmann Entropy can be calculated and which fulfil the

Equal Accessibility Assumption: there are many highly structured 'clumps' of particles (e.g., humans, chairs, planets), which seems to speak against the possibility of the universe being subject to the statistical second law. However, two arguments can be made here that render the existence of such structures less problematic: (i) we have seen in the example in Section 3.4 that statistical entropy is not a measure of visible disorder and that structured states can have higher entropy than visibly disordered states; (ii) even if we assume that obviously structures are low-entropy regions in the universe, one could argue that they are normally temporary (e.g., humans, chairs, and planets all have finite lifespans), and their creation is associated with an entropy increase elsewhere (e.g., the creation of humans, chairs, and planets requires energy, whose generation can be associated with an entropy increase elsewhere). With respect to argument (i), Denbigh (1989, p. 329) writes:

> In my view some of the many discussions in the literature on the evolution of the universe from the Big Bang onwards have been weakened by attempts to apply the notions of 'chaos' and 'disorder' – and also 'uniformity' – as if these were equivalent to using the Second Law.

Given the discussion in Section 3.4, where I argued that statistical entropy is not a measure for visible disorder, I agree with Denbigh (1989). The fact that there are visibly ordered regions in the universe likely tells us very little about the statistical entropy of those regions and should therefore not be seen as indicative of the entropy development of the universe. However, I also argued that statistical entropy tracks a more sophisticated notion of disorder, and the calculation of entropy should involve a detailed consideration of which degrees of freedom need to be taken into account for each phenomenon. This would indicate that computing the statistical entropy of the whole universe would be fiendishly difficult and that treating the universe as a system of identical, non-interacting, elastic point-particles would not be appropriate. There is clearly currently no way to formally perform this calculation, so that the phrase 'entropy of the universe' might be, as Planck (1897, cited in Uffink, 2003) puts it, meaningless. This judgement has been echoed in more recent works as well (Albert, 2000, Chapters 4–5; Wald, 2006), with which I am inclined to agree, given my own argumentation earlier.

In summary, the verdict on the veracity of the Cosmic Entropy Assumption has not yet been delivered, although, in the round, philosophers and physicists are sceptical about its veracity. However, what seems to be clear from the preceding considerations is that, even if the entropy of the universe can be defined and it is subject to the statistical second law, there is little prospect of us assigning a numerical value to the entropy of the universe, much less

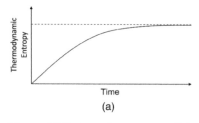
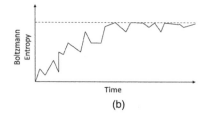

Figure 10 Entropy increase predicted by (i) the empirical second law and (ii) the statistical second law.

continuously use such a value to 'date' different stages of the universe. Accordingly, the actual assignments of time-orderings would have to be made based on more easily observable sub-processes and regions of the universe. We will return to the consequences of this realisation in what follows.

The second assumption we need to make to link the Universal Statistical Arrow of Time to an actual increase in the Boltzmann Entropy of the universe is that there were suitable initial conditions for the statistical second law (Section 4.2) to predict an increase in entropy, rather than constant maximum-entropy. In particular, given that high-entropy macrostates are actually over-whelmingly more likely (based on the number of microstates that can instantiate them, Section 3.2), the most likely development for any system is to stay close to a high-entropy macrostate. Accordingly, we would only see an overall increase in the statistical entropy of the universe if its initial macrostate was a low-entropy one (Figure 10). This assumption is known as the Past Hypothesis (e.g., Albert, 2000, p. 96; North, 2011, pp. 12–16):

Past Hypothesis: The initial macrostate of the universe was a low-entropy state.

The plausibility of the Past Hypothesis, like the one of the Cosmic Entropy Assumption, depends on the underlying model for the universe's origin and early development. It is therefore contingent on the resolution of fundamental debates in cosmology and, therefore, currently undecided. It is notable that for one of the most accepted models of the early universe, the Big Bang Model, an argument can be made for a low-entropy initial macrostate (North, 2011, p. 14):

> Although this has not been worked out rigorously, there is a rough answer that strikes many people as plausible. Immediately after the big bang, the universe was in a uniformly hot "soup," with matter and energy uniformly distributed in thermal equilibrium. This state did have high Thermodynamic Entropy. The thought is that it had extremely low entropy due to gravity. Gravity is an

attractive force: matter tends to clump up under this force, and then to stay clumped up. We know from thermodynamics that maximal entropy states are the equilibrium states toward which systems tend to evolve and then stay. For systems primarily under the influence of gravity, then, a clumped-up state has high entropy. The early state of the universe, non-clumped-up and uniformly spread out, had extremely low entropy due to gravity.

The main tenet of this argument is the identification of high-entropy states with equilibrium states, similar to Brown and Uffink's (2001) proposal that the notion of entropy can be reduced to that of equilibrium. In a high-gravity situation as hypothesised by the Big Bang model, the equilibrium state is one of gravitational clusters ('clumps') and the proposed uniform state is therefore an unstable, non-equilibrium, low-entropy state. Notably, this argument is based on another example of the kind of phenomenon discussed in Section 3.4, where entropy does not track visible disorder. However, as North (2011, quotation above) points out, this argument has not been developed into a formal model yet and is also not universally accepted.

Price (1997, pp. 32–7), returning to an argument made by Boltzmann (1896/ 1964), has proposed an alternative scenario that would lead to an increase in statistical entropy for the observable universe but does not require the adoption of the Past Hypothesis. He proposes that the development of the universe's entropy actually consists of small fluctuations around a high-entropy state, which we would expect from the definition of statistical entropy (Section 3.2). However, some time before the existence of humankind, the universe experienced a relatively large one of those expected fluctuations downward from its typical high-entropy state. This very fluctuation made it possible for our structured world (containing humans, chairs, planets etc.) to come into being. Furthermore, the increase in entropy we are currently observing in the world is not a trend in the long-term development of the universe's entropy but the temporary reversal of this fluctuation (Figure 11). Price (1997, p. 47) concludes that the statistical second law is insufficient as a grounding for the Arrow of

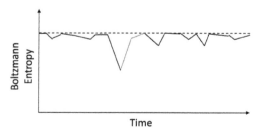

Figure 11 Price's (1997) scenario of a local fluctuation leading to an observed, but atypical, entropy increase. The universe is currently in the red region of entropy increase.

Time since it is likely that the long-term development of the universe will include both periods of decreasing and of increasing Boltzmann Entropy. While it is unlikely that humankind would be able to exist in a universe at maximum entropy, the fact that our collective lifespan will be characterised by increasing entropy is not a sufficient reason to define the direction of time as this is likely atypical for the universe as a whole.

In an extension of Price's (1997) thesis, the existence of entropy fluctuations in the statistical second law (Section 4.4) is clearly problematic if we want to assert that the entropy development of the universe tracks the direction of time deterministically. The conclusions we could draw from a hypothetical film of the entropy of the universe would simply be that it is much (much!) more likely that a low-entropy state happens before a high-entropy state. However, we would not be able to rule out that the sequence of entropy states had been recorded during one of the unlikely, but possible, downward fluctuations in entropy (Figure 9). Accordingly, in contrast to the Empirical Arrow of Time (Section 5.1), the Universal Statistical Arrow of Time only probabilistically fulfils the **definitional component** of an entropy-grounding.

What is the explanatory potential of the Universal Statistical Arrow of Time for the constituting-processes of our experience of the direction of time? In order to fulfil the **explanatory component** of a grounding in the statistical second law, we would need to show that an increase in the entropy of the universe implies that processes like aging, memory-formation, causality, and others only happen in one direction, namely, are irreversible. The standard argument for this implication requires positing that (i) those processes are individually entropy increasing and (ii) that the overall increase in entropy predicted by the statistical second law in conjunction with the Cosmic Entropy Assumption and the Past Hypothesis requires (in a strong statistical sense; see previous discussion) those individual processes to be entropy increases and is therefore explanatory for their directionality. However, as North (2011, p. 15) states:

This is not a rigorous argument. It is a plausibility claim that the theory should be able to ground our records in this way, given Boltzmann's reasoning in statistical mechanics, and given big bang cosmology's account of the formation of stars and galaxies, which in turn lead to the existence of beaches and people, who in turn lead to the existence of frozen popsicles, and so on.

In summary, the Universal Statistical Arrow of Time has the potential to explain our experience of time but there currently exists no formal, sufficiently detailed, causal argument to actually do so. Progress on this matter would require significant advances in our understanding of both the development of the universe as well as the workings of the constituting-processes of our experience of time.

It should be apparent from the preceding discussion that many contingencies on the Universal Statistical Arrow of Time – in particular, the Cosmic Entropy Assumption and the Past Hypothesis – result from the fact that it is a grounding in the Statistical Entropy of the universe itself, that is, from the domain of application. In the next section, we will discuss an alternative entropy-grounding that focuses on our local experience of time instead.

5.3 The Local Statistical Arrow of Time

It is possible to decouple the grounding of the Arrow of Time in statistical entropy from any assertions about the entropy of the universe as such: namely, by requiring that the statistical second law (Section 4.2) grounds our experience of the direction of time rather than time in an abstract, non-experiential sense (Sklar, 1993, p. 390):

Local Statistical Arrow of Time: The Local Statistical Arrow of Time points in the direction of the average increase in entropy during the constituting processes of our experience of time (e.g., aging, memory formation, causality), that is, we can explain the asymmetry in those processes by reference to the statistical second law.

Given that we are now only concerned with our experience of the direction of time (which is immediately accessible to us), the Local Statistical Arrow of Time has lost its definitional function. To return to the example of watching a film and trying to time-order different frames from this film, we would have to try to identify a constitution-process of our experience of time to do so, for example, by watching out for cause-and-effect pairs or explicit accounts of memory formation. It therefore does not fulfil the ***definitional component*** of a grounding in statistical entropy.

However, the Local Statistical Arrow of Time could have an explanatory function, that is, it has the potential to fulfil the ***explanatory component*** of a grounding. We can now explain the fact that our experience of time is asymmetrical by referring to the fact that the processes that constitute this experience are governed by the asymmetrical statistical second law. While the Local Statistical Arrow of Time avoids the additional assumptions about the nature of the universe that we had to make in the definition of the Universal Statistical Arrow of Time (Section 5.2), to be truly explanatory of our experience of time, we need to make an assumption about the nature of the processes underlying our experience of time:

Local Reduction Assumption: (i) All local processes that govern our experience of time are reducible to statistical mechanical systems, namely, to particle dynamics; (ii) a successful reduction of those processes confirms that they take place in systems to which the statistical second law is applicable; and (iii) the initial state of each such process is of lower entropy than the final state.

The first requirement of the Local Reduction Assumption, the possibility of reducing those processes to the interactions of ensembles of particles, is unlikely to be controversial: it is a well-established fact that, on a microscopic level, the world is constituted of atoms and their interactions.

However, the second requirement, that those processes will be governed by the statistical second law, is much harder to ascertain. In particular, we currently have an imperfect understanding of even the macroscopic, neurological and physical, processes that underpin experiences like memory-formation and aging. There is therefore no easy 'lever' available to even begin a reduction of those systems to statistics of particle interactions. Even at this stage it is apparent, however, that both those biological processes and causal processes are not easily framed as taking place in closed systems: for example, in the case of memory-formation, over the lifespan of a human, there is a complex exchange between energy drawn from the environment through metabolic processes and the maintenance of neurological functions. Similarly, it is unlikely that the instantiation of the Equal Accessibility Assumption, which still needs to be fulfilled in those systems for the second law to be applicable, would take the form of the *Stosszahlansatz* that was derived for and tested on systems of identical, non-interacting, elastic point-particles. Instead, it is much more likely that those phenomena are similar to the systems of rod-like particles discussed in Section 3.4.2 and that the Equal Accessibility Assumption needs to take a different form to assure that micro-states on different micro-scales are equally accessible. It now almost goes without saying that we should also not expect those processes to align with any transition from visible order to disorder.

The third requirement of the Local Reduction Assumption is based on the fact that, in order for the statistical second law to predict an overall increase rather than fluctuations around the maximum entropy value, the initial state of each of those processes would have to be at a lower entropy value than the final state. This is not quite as severe an assumption as the Past Hypothesis, as the low-entropy state only has to obtain for each process and could actually be different each time the process runs, for example, one could imagine that the entropy of the human brain increases stepwise through each memory formation process, with each such process starting from a slightly higher level than before.

In addition, such a reduction would run into the same fundamental problems that the project of reducing thermodynamics to statistical mechanics has already encountered (Section 3.1): it would have to account for the translation of probabilistic predictions, as made by the statistical second law, into outcomes of (at least seemingly) deterministic processes. This appears to be a particular problem with respect to causation, which in its most paradigmatic instances (e.g., physical impact) seems very clearly to be deterministic. That is, even a vanishingly small probability of, say, Newton's Cradle not resulting in an impact on the last pendulum in the sequence would strongly contradict our intuitions about the world.

An alternative approach to a straightforward physical reduction would be to attempt a reduction of both the statistical second law and the constituting-processes of our experience of time to a third, even more fundamental quantity. Sklar (1993, pp. 387–411) reviews several such attempts, including a reduction to information and information exchanges. This latter approach was popular in the second half of the twentieth century (e.g., Brillouin, 1951; Bennett, 1973; 1982). However, the information-theoretic approach to entropy is currently more controversially debated (for arguments against, see, e.g.: Earman & Norton 1998; 1999; Norton, 2005; for arguments for: Bub, 2001. For our purposes, it is sufficient to note that no full reduction to any third quantity has yet been achieved. Therefore, the explanatory potential of the Local Statistical Arrow of Time remains exactly this: a potentiality of offering an explanation that would avoid making assumptions about the nature of the universe (in contrast to the Universal Statistical Arrow of Time, Section 5.2), but for which currently no straightforward route to formalisation has been identified.

5.4 Conclusion

It is easily apparent that the derivation of the two Statistical Arrows of Time from the statistical second law requires the introduction of more additional assumptions than does the derivation of the Empirical Arrow of Time. The Global Statistical Arrow of Time requires two additional assumptions (Section 5.2): the Cosmic Entropy Assumption and the Past Hypothesis. The Local Statistical Arrow of Time requires one additional assumption: the Local Reduction Assumption (Section 5.3). Carrying over from the prior derivation of the statistical second law is the Equal Accessibility Assumption (Section 4.3). Therefore, the validity of the Statistical Arrows of Time depends on a comparatively large number of contingencies. Furthermore, as demonstrated in the discussions of each assumption above, the assumptions are not trivially fulfilled and decisions about their veracity depend on advances on other, highly contested, questions in science and philosophy: in the case of the Cosmic Entropy Assumption and the Past Hypothesis on advances in our understanding of the universe's cosmological development; in the case of the Local Reduction Assumption on a better understanding of how atomic and molecular phenomena underpin biological, neurological, and causal processes. I think it is fair to say that a resolution of those questions is currently not in sight, and that it would be overly optimistic to assume that any such resolution would make those assumptions true rather than false. There are therefore several authors who contest the suitability of the Global Statistical Arrow of Time or the Local Statistical Arrow of Time as a (sole) definition for the direction of time (e.g., Sklar, 1993; Albert, 2000).

In addition, even if one is optimistic about the validity of the assumptions that the derivations of the Statistical Arrows of Time introduce, the statistical nature of those groundings means that their definitional powers, defined earlier as establishing a time-ordering of events, is reduced. As Price (1997) points out, the fact that higher-entropy states are overwhelmingly likely to follow lower-entropy states (and not the other way around) could be a temporary state of the universe's development (Section 5.2). Even if a general entropy increase is assumed, that is, we are not in a 'fluke' situation as hypothesised by Price (1997), it could be the case that the events we are trying to time-order based on am Statistical Arrow of Time are part of an extremely unlikely, but possible, high-to-low entropy development. Accordingly, the Statistical Arrows of Time only provide a most likely time-ordering, not a deterministic definition of the direction of time.

6 Conclusions

In Section 1, we defined two aims for this Element:

Aim 1: Reconstructing, analysing, and comparing different derivational routes to a grounding of the Arrow of Time in entropy.

Aim 2: Evaluating the link between entropy and visible disorder, and the related claim of an alignment of the Arrow of Time with a development from order to visible disorder.

In the following, I will first discuss the insights we obtained through fulfilling Aim 1 (Section 6.1) and then summarise the results of the evaluation of the link between entropy and visible disorder we undertook (Section 6.2).

6.1 Entropy-Groundings of the Arrow of Time

In the last four sections, we have traced the derivational routes of different entropy-groundings of the Arrow of Time. Comparing Figure 12 to Figure 1, we can see that the analysis in this Element has led to an unpacking of three different derivational routes for entropy-groundings of the Arrow of Time, and we have now gained a better understanding of the contingencies that each route introduces. Based on this analysis, we are therefore able to assess the relative epistemic benefits of each entropy-grounding: (i) the Empirical Arrow of Time fulfilled the definitional component of a grounding, but has no explanatory potential or power (Section 5.1); (ii) the Universal Statistical Arrow of Time has the potential to fulfil both the definitional and explanatory components of a grounding, but a realisation of this potential would require the resolution of longstanding debates on the veracity of the three assumptions it depends on (Section 5.2); (iii) the Local Statistical Arrow of Time does not fulfil the definitional component of a grounding, but has the potential to fulfil the explanatory component if longstanding debates on the veracity of the two assumptions it depends on are resolved.

What are the overarching conclusions we can draw from the results of this analysis? I maintain that there are three general conclusions that this comprehensive analysis of the different derivational routes to entropy-groundings of the Arrow of Time entails. Those will be discussed in more detail in what follows.

Conclusion 1: Entropy-groundings of the Arrow of Time are highly dependent notions.

It is immediately obvious from the coexistence of those different derivational routes (Figure 12) that the precise formulation and the epistemic function of different entropy-groundings of the Arrow of Time depend on the choice of

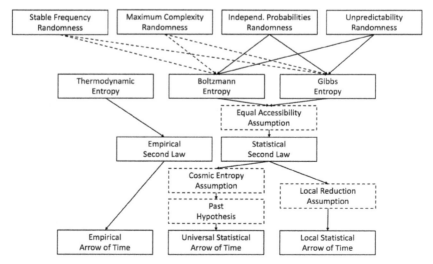

Figure 12 Root-concepts of and derivational routes to entropy-groundings of the Arrow of Time.

ancestor concepts and of derivational method. Therefore, it is not possible to view any entropy-grounding as a notion that 'can be read off' from the world. Instead, each grounding of the direction of time is derived from multiple ancestor notions. Entropy-groundings of the Arrow of Time are therefore deeply rooted in the foundational framework of statistical mechanics and thermodynamics. On the one hand, this means that unresolved questions about the ancestor concepts – for example, about the correct definition of entropy; the relationship between entropy and disorder (also see Section 6.2); the correct derivation of the second law of thermodynamics, the reduction of thermodynamics to statistical physics, and so on – influence the epistemic functions of different entropy groundings. On the other hand, the dependency on fundamental entities means that advances on the foundations of statistical physics and thermodynamics will likely improve the fulfilment of both the definitional and explanatory components of each grounding (see also Conclusion 3).

Conclusion 2: There is currently no entropy-grounding of the Arrow of Time that is both definitional and explanatory for the direction of time.

One of the most prominent results of this analysis is the lack of a clearly distinguished 'best' candidate among the three different entropy-groundings. None of the groundings fulfils both the explanatory and the definitional component of a grounding, and the one that has the potential to do so, the Universal Statistical Arrow of Time, is the one whose derivation introduces the most contingencies.

Conclusion 2 could lead one to think that the project of deriving an entropy-grounding of the Arrow of Time has failed and that one should turn to other physical quantities to derive a grounding for the direction of time. Indeed, some authors have come to this conclusion (prominently, Price, 1997) and there exist alternative proposals to ground the Arrow of Time in electromagnetic radiation (Frisch, 2006); in complexity (Lineweaver *et al.*, 2013); and in weak interaction (Golosz, 2017; Roberts, 2022). Each of those proposals deserves a detailed analysis of its own to decide on its relative merits and demerits compared to entropy-groundings; I therefore remain equivocal about the possibility that one of those groundings will turn out to have better explanatory and definitional properties than the three entropy-groundings discussed here. However, by making clear the derivational routes to and respective advantages/disadvantages of each entropy-grounding of the Arrow of Time, I hope that this Element has paved the way for a truly comparative discussion of different groundings.

Furthermore, rather than abandon the project of deriving an entropy-grounding of the Arrow of Time, I view Conclusion 2 as additional motivation to 'work very hard on statistical mechanics' (Earman & Norton, 1998, p. 440), that is, as motivation to work towards resolving the foundational debates about the validity of the assumptions that go into the derivation of different entropy-groundings, and that currently impede the realisation of their explanatory and definitional potential.

Conclusion 3: The lack of a fully definitional and explanatory entropy-grounding of the Arrow of Time highlights the importance of reduction- and unification-projects in statistical mechanics.

Throughout the analysis in this Element, it has become apparent that a fully satisfactory entropy-grounding of the Arrow of Time would require the resolution of some pertinent and longstanding questions on the reduction of a variety of processes to particle dynamics. This is particularly apparent in the case of the Cosmic Entropy Assumption (Section 5.2) and the Local Reduction Assumption (Section 5.3), whose veracity crucially depends on how entropy-relevant, macroscopic processes are instantiated on the level of molecules and atomic particles. A furthering of the general reduction-project in science and statistical mechanics would, therefore, likely be the most efficient way of improving the explanatory power of the two Statistical Arrows of Time.

Furthermore, the Universal Statistical Arrow of Time, which has the potential to fulfil the definitional and explanatory components of grounding for the direction of time, requires the resolution of open questions in cosmology to decide on the veracity of the Cosmic Entropy Assumption and the Past Hypothesis (Section 5.2). Similarly, gaining a better understanding of the neurological and biological processes that underpin aging and memory formation will be required to decide on the veracity of the Local Reduction Assumption. As such, the analysis of the different derivational routes of the three entropy-groundings of the Arrow of Time does not just emphasise the importance of intertheoretical reduction, but of a unification and close integration of other areas of science with statistical physics and thermodynamics.

6.2 Entropy and Disorder

Throughout the Element, we discussed the question of whether statistical entropy is a measure of visible disorder (formalised as randomness, Section 2) and whether the Arrow of Time, if grounded in statistical entropy, correspondingly also points in the direction of increasing disorder. We have seen (Section 1.1) that this supposed link between an increase in entropy and disorder is often used in conceptual arguments about the direction of a variety of processes.

However, in Section 3.4, we analysed a counterexample to the claim that statistical entropy is a measure for visible disorder and demonstrated that, for most phenomena beyond a class of very simplified particle systems, it is unlikely that high-entropy states will coincide with visible disorder.

Conclusion 4: Statistical entropy is not a measure for visible disorder.

I have also argued that statistical entropy can be seen as a measure for a more abstract notion of disorder (Section 3.4.3): a deviation from a perfectly ordered state defined by the choice of phase space and, for coarse-grained entropies, partition. Thereby, the phase space and partition one uses is not arbitrary but contingent on the properties of the system under consideration. However, the severing of the close connection between high entropy and visible disorder that arguments based on the assumption that processes will naturally move from order to visible disorder should be avoided. Similarly, one should be careful in engaging in the reverse kind of reasoning, that is, assuming that a macrostate that looks visibly disordered has high entropy and that ordered states always have low entropy.

References

Albert, D. Z. (2000). *Time and Chance*. Cambridge, MA: Harvard University Press.

Aoki, I. (2012). *Entropy Principle for the Development of Complex Biotic Systems: Organisms, Ecosystems, the Earth*. London: Elsevier.

Bennett, C. H. (1973). Local Reversibility of Computation. *IBM Journal of Research and Development* **17**, pp. 525–32.

Bennett, C. H. (1982). The Thermodynamics of Computation: A Review. *International Journal of Theoretical Physics* **21**, pp. 334–7.

Boltzmann, L. (1872). Weitere Studien ueber Waermegleichgewicht unter Gasmolekuelen. *Wiener Berichte* **96**, pp. 275–370.

Boltzmann, L. (1896/1964). *Lectures on Gas Theory*. New York: Dover.

Brillouin, L. (1951). Maxwell's Demon Cannot Operate: Information and Entropy I. *Journal of Applied Physics* **22**, pp. 334–7.

Brown, H. R. & Uffink, J. (2001). The Origins of Time-Asymmetry in Thermodynamics: The Minus-First Law. *Studies in History and Philosophy of Modern Physics* **32**, pp. 525–38.

Brown, H. R., Myrvold, W., & Uffink, J. (2009). Boltzmann's H-Theorem, Its Discontents, and the Birth of Statistical Mechanics. *Studies in History and Philosophy of Modern Physics* **40**, pp. 174–91.

Bub, J. (2001). Maxwell's Demon and the Thermodynamics of Computation. *Studies in History and Philosophy of Modern Physics* **4**, pp. 569–79.

Burgers, J. M. (1970). Entropy and Disorder. *British Journal for the Philosophy of Science* **5**, pp. 70–1.

Callender, C. (1999). Reducing Thermodynamics to Statistical Mechanics: The Case of Entropy. *The Journal of Philosophy* **96**, pp. 348–73.

Church, A. (1940). On the Concept of a Random Sequence. *Bulletin of the American Mathematical Society* **46**, pp. 130–5.

Coffa, J. A. (1972). Randomness and Knowledge. *PSA: Proceedings of the Biennial Meeting of the Philosophy of Science Association* **1972**, pp. 103–15.

Correia, F. & Schnieder, B. (2012, Eds.). *Metaphysical Grounding: Understanding the Structure of Reality*. Cambridge: Cambridge University Press.

Copeland, B. J. & Proudfoot, D. (2005). Turing and the Computer. In Copeland, B. J. (Ed.), *Alan Turing's Automatic Computing Engine: The Master Codebreaker's Struggle to Build the Modern Computer*. Oxford: Oxford University Press, pp. 107–48.

DeMol, L. (2021). Turing Machines. In Zalta, E. (Ed.). *The Stanford Encyclopedia of Philosophy* (Winter 2021 Edition), https://plato.stanford.edu/archives/win2021/entries/turing-machine.

Denbigh, K. G. (1989). Note on Entropy, Disorder and Disorganisation. *British Journal for the Philosophy of Science* **40**, pp. 323–32.

Eagle, A. (2005). Randomness Is Unpredictability. *British Journal for the Philosophy of Science* **56**, pp. 749–90.

Earman, J. & Norton, J. D. (1998). Exorcist XIV: The Wrath of Maxwell's Demon. Part I. From Maxwell to Szilard. *Studies in History and Philosophy of Modern Physics* **29**, pp. 435–71.

Earman, J. & Norton, J. D. (1999). Exorcist XIV: The Wrath of Maxwell's Demon. Part II. From Szilard to Landauer and Beyond. *Studies in History and Philosophy of Modern Physics* **30**, pp. 1–40.

Ehrenfest, P. & Ehrenfest, T. (1907). Ueber Zwei Bekannte Einwende Gegen das Boltzmannsche H-Theorem. *Physikalische Zeitung* **8**, pp. 311–14.

Ehrenfest, P. & Ehrenfest, T. (1909). *Begriffliche Grundlagen der Statistischen Auffassung in der Mechanik*. Leipzig: Teubner.

Frenkel, D. (1999). Entropy-Driven Phase Transitions. *Physica A* **263**, pp. 26–38.

Frigg, R., & Werndl, C. (2011). Entropy: A Guide for the Perplexed. In Beisbart, C., & Hartmann, S. (Ed.). *Probabilities in Physics*. Oxford: Oxford University Press.

Frisch, M. (2006). A Tale of Two Arrows. *Studies in the History and Philosophy of Modern Physics* **37**, pp. 542–58.

Georgii, H.-O. & Zagrebnov, V. (2011). Entropy-Driven Phase Transitions in Multitype Lattice Gas Models. *Journal of Statistical Physics* **102**, pp. 35–67.

Gibbs, J. W. (1902/1960). *Elementary Principles in Statistical Mechanics*. New York: Dover.

Gobbo, D., Ballone, P., & Garabato, B. D. (2020). Coarse-Grained Model of Entropy-Driven Demixing. *The Journal of Physical Chemistry B* **124**, pp. 9267–74.

Golosz, J. (2017). Weak Interactions: Asymmetry of Time or Asymmetry in Time? *Journal for General Philosophy of Science* **48**, pp. 19–33.

Goodman, N. (1955/1983). *Fact, Fiction and Forecast*. Cambridge, MA: Harvard University Press.

Haglund, J. (2017). Good Use of a Bad Metaphor: Entropy as Disorder. *Science & Education* **26**, pp. 205–14.

Hempel, C. G. (1945). Studies in the Logic of Confirmation. *Mind* **54**, pp. 1–26, 97–121.

Hershey, D. (2009). *Entropy Theory of Aging Systems: Humans, Corporations and the Universe*. London: Imperial College Press.

Kalman, R. E. (1994). Randomness Reexamined. *Modelling, Identification and Control* **15**, pp. 141–51.

Kolmogorov, A. N. & Uspenskii, V. A. (1988). Algorithms and Randomness. *SIAM Theory and Probability of Applications* **32**, pp. 389–412.

Landsman, K. (2020). Randomness? What Randomness? *Foundations of Physics* **50**, pp. 61–104.

Laplace, P.-S. (1826/1851). *Philosophical Essay on Probabilities*. New York: Dover.

Lavis, D. A. (2005). Boltzmann and Gibbs: An Attempted Reconciliation. *Studies in History and Philosophy of Modern Physics* **36**, pp. 245–73.

Le Poidevin, R., & MacBeath, M. (1993, Eds.). *The Philosophy of Time*. Oxford: Oxford University Press.

Leff, H. S. (2007). Entropy, Its Language and Interpretation. *Foundations of Physics* **37**, pp. 1744–66.

Lineweaver, C. H., Davies, P. C. W., & Ruse, M. (2013, Eds.). *Complexity and the Arrow of Time*. Cambridge: Cambridge University Press.

Loschmidt, J. (1876). Ueber den Zustand des Waermegleichgewichts eines Systemes von Koerpern mit Ruecksicht auf die Schwerkraft. *Sitzungsberichte der Akademie der Wissenschaften zu Wien, Mathematisch-Naturwissenschaftlichle Klasse* **73**, pp. 128–42.

Margenstern, M. (2000). Frontier between Decidability and Undecidability: A Survey. *Theoretical Computer Science* **231**, pp. 217–51.

Martin-Loef, P. (1966). The Definition of a Random Sequence. *Information and Control* **9**, pp. 602–19.

Maxwell, J. C. (1867). Letter to P. G. Tait, 11/12/1867. In Knott, C. G. (1911). *Life and Scientific Work of Peter Guthrie Tait*. Cambridge: Cambridge University Press, p. 214.

McTaggart, J. M. E. (1908). The Unreality of Time. *Mind* **17**, pp. 457–74.

Mellor, D. H. (1998). *Real Time II*. London: Routledge.

North, J. (2011). Time in Thermodynamics. In Callender, C. (Ed.). *The Oxford Handbook of Time*. Oxford: Oxford University Press, pp. 1–46.

Norton, J. D. (2005). Eater of the Lotus: Landauer's Principle and the Return of Maxwell's Demon. *Studies in History and Philosophy of Modern Physics* **36**, pp. 375–411.

Norton, J. D. (2013). All Shook Up: Fluctuations, Maxwell's Demon and the Thermodynamics of Computation. *Entropy* **2013**, pp. 4432–83.

Poincaré, H. (1890). Sur le Probleme des Trois Corps et les Equations de la Dyanmiqué. *Acta Mathematica* **13**, pp. 1–270.

Popper, K. R. (1959/2002). *The Logic of Scientific Discovery*. London: Routledge.

Price, H. (1997). *Time's Arrow and Archimedes' Point: New Directions for the Physics of Time*. Oxford: Oxford University Press.

Prior, A. N. (1967). *Past, Present, Future*. Oxford: Oxford University Press.

Rickles, D. (2016). *The Philosophy of Physics*. Cambridge: Polity

Roberts, W. B. (2022). *Reversing the Arrow of Time*. Cambridge: Cambridge University Press.

Robertson, K. (2020). In Search of the Holy Grail: How to Reduce the Second Law of Thermodynamics. Forthcoming in *The British Journal for the Philosophy of Science*.

Sklar, L. (1993). *Physics and Chance: Philosophical Issues in the Foundations of Statistical Mechanics*. Cambridge: Cambridge University Press.

Smoluchowski, M. (1912). Experimentell nachweisbare, der ueblichen Thermodynamik wiedersprechende Molekularphenomaene, *Physikalische Zeitschrift* **13**, pp. 1069–80.

Smoluchowski, M. (1914). Gueltigkeitsgrenzen des Zweiten Hauptsatzes der Waermetheorie. In *Vortraege ueber die Kinetische Theorie der Materie und der Elektrizitaet*. Leibzig: Teuber.

Szilard, L. (1929/1972). On the Decrease of Entropy in a Thermodynamic System by Intervention of Intelligent Beings. In *The Collected Works of Leo Szilard: Scientific Papers*. Boston: Massachusetts Institute of Technology Press.

Turing, A. M. (1936). On Computable Numbers, with an Application to the Entscheidungsproblem. *Proceedings of the London Mathematical Society* **42**, pp. 230–65.

Uffink, J. (2003). Irreversibility and the Second Law of Thermodynamics. In Greven, A., Keller, G., & Warnecke, G. (Eds.). *Entropy*. Princeton, NJ: Princeton University Press, pp. 121–46.

Van Rield, R. & van Gulick, R. (2019). Scientific Reduction. In Zalta, E. (Ed.). *The Stanford Encyclopedia of Philosophy* (Spring 2019 Edition), https://plato.stanford.edu/archives/spr2019/entries/scientific-reduction/.

Von Mises, R. (1957). *Probability, Statistics and Truth*. New York: Dover.

Wald, R. M. (2006). The Arrow of Time and the Initial Conditions of the Universe. *Studies in History and Philosophy of Modern Physics* **37**, pp. 394–8.

Weinberg, S. (2008). *Cosmology*. Oxford: Oxford University Press.

Werndl, C. (2009a). Are Deterministic Descriptions and Indeterministic Descriptions Observationally Equivalent? *Studies in History and Philosophy of Modern Physics* **40**, pp. 232–42.

Werndl, C. (2009b). What Are the New Implications of Chaos for Unpredictability? *British Journal for the Philosophy of Science* **60**, pp. 195–220.

Werndl, C. (2011). On the Observational Equivalence of Continuous-Time Deterministic and Indeterministic Descriptions. *European Journal for Philosophy of Science* **1**, pp. 193–225.

Zuchowski, L. C. (2012). Disentangling Complexity from Randomness and Chaos. *Entropy* **14**, pp. 177–212.

Zuchowski, L. C. (2017). *A Philosophical Analysis of Chaos Theory*. London. Palgrave Macmillan.

Cambridge Elements ⁼

The Philosophy of Physics

James Owen Weatherall

University of California, Irvine

James Owen Weatherall is Professor of Logic and Philosophy of Science at the University of California, Irvine. He is the author, with Cailin O'Connor, of *The Misinformation Age: How False Beliefs Spread* (Yale, 2019), which was selected as a *New York Times* Editors' Choice and Recommended Reading by *Scientific American*. His previous books were *Void: The Strange Physics of Nothing* (Yale, 2016) and the *New York Times* bestseller *The Physics of Wall Street: A Brief History of Predicting the Unpredictable* (Houghton Mifflin Harcourt, 2013). He has published approximately fifty peer-reviewed research articles in journals in leading physics and philosophy of science journals and has delivered over 100 invited academic talks and public lectures.

About the Series

This Cambridge Elements series provides concise and structured introductions to all the central topics in the philosophy of physics. The Elements in the series are written by distinguished senior scholars and bright junior scholars with relevant expertise, producing balanced, comprehensive coverage of multiple perspectives in the philosophy of physics.

Cambridge Elements ☰

The Philosophy of Physics

Elements in the Series

A full series listing is available at: www.cambridge.org/EPPH